Surface
Ice
Rescue

Surface
Ice
Rescue

Walt "Butch" Hendrick
Andrea Zaferes

Fire Engineering®
BOOKS & VIDEOS

Copyright © 1999 by
Fire Engineering Books & Videos
110 S. Hartford Ave., Ste 200
Tulsa, Oklahoma 74120 USA

800.752.9764
+1.918.831.9421
info@fireengineeringbooks.com
www.FireEngineeringBooks.com

Senior Vice President: Eric Schlett
Operations Manager: Holly Fournier
Sales Manager: Josh Neal
Managing Editor: Mark Haugh
Production Manager: Tony Quinn
Developmental Editor: Chris Barton
Cover Designer: Steve Hetzel
Book Designer: Pat Rasch

Library of Congress Cataloging-in-Publication Data

Hendrick, Walt, 1947-
Surface ice rescue / Walt "Butch" Hendrick, Andrea Zaferes.
   p. cm.
ISBN: 0-912212-85-3
ISBN13: 978-0-912212-85-2
Includes bibliographical references and index.
1. Hypothermia.   2. Ice accidents.   3. Rescue work.
I. Zaferes, Andrea, 1965- .   II. Title.
RA88.5.H46 1999
616.9'89—dc21   99-14116
                                                                              CIP

Printed in the United States of America

13 14 15 16   26 25 24 23

About the Authors

Walt "Butch" Hendrick has been training public safety dive teams and rescue personnel for more than twenty-five years in more than fifteen countries. Founder and president of Lifeguard Systems, Inc., he is a major innovator in and contributor to the water rescue industry. Many water and ice accident victims around the world are alive today because of his innovations, training programs, publications, audiovisual materials, equipment designs, public speaking, and actual rescue efforts.

Andrea Zaferes began teaching diving with Dr. Lee Somers and Karl Huggins at the University of Michigan's Scientific Diving Program. She served as a diving safety officer for the American Museum of Natural History's Animal Behavior Research Department and had three research papers published by the age of twenty-two. She took her first diving rescue course at age sixteen. Andrea is vice president and program director of Lifeguard Systems, Inc., and has been teaching water rescue for more than ten years.

Acknowledgments

This book is dedicated to the more than 20,000 fire, police, EMS, military, Coast Guard, and dive team personnel who have allowed us to train them in surface and subsurface rescue over the past twenty-five years. We also thank the hundreds of surface ice rescue students who have proved that performing a complete surface rescue in less than three minutes is very possible. Thank you. You have taught us so much.

Special thanks to those who have given us so much of their time and support, and for their caring for others: Orlando Abreus; Ken Balfrey; Adam Cadan; John Earl; Michael Emmerman; Darlene Esposito, D.C.; Lt. Todd Hanson; David Harrison; Capt. Patrick Kilbride; Bob Leambruno; David McCoy; Dave Mandel; Mike Mulligan; Wendy Novack; Andy Schmidt; Kurt Semmel; Karen Van Hoesen, M.D.; Dan Vircik; and Jason Yates.

The warmest of thanks to George Safirowski, a steady and wonderful friend, who can never be replaced.

Thanks to Bob Davis for all of his knowledge, experience, willingness, and ability to build us safer, more effective ice rescue equipment, based on what rescuers ask for and need.

And thanks to Ralph Dodds and Craig Nelson for their help in completing so many projects, including this one.

Table of Contents

Preface

Two girls who have fallen through the ice are observed by a man walking his dog. The man makes the poor decision to go out onto the ice to attempt to help them. In doing so, he falls through. The members of the responding fire department have little or no ice rescue training and no ice rescue equipment. They find the girls hanging on to supportive ice in an alert condition, and the man is bobbing up and down, submerging each time. The rescuers don't don personal flotation devices, nor do they don gloves or other personal protective equipment. Since the weather is relatively warm for a winter day, some of them are wearing nothing more than T-shirts and jeans. Several are wearing turnout gear.

Obviously they have no protocol for surface ice rescue and no immediate plan. The responders find a metal canoe, and a few of them jump into it and begin breaking up the ice with an oar and bare hands to get to the man, who is obviously in the worst condition. No communication to the victims is established. No one draws a profile map, and no one marks the shore in front of each victim. A medevac helicopter arrives, and the crew members are led to the scene by a firefighter. They are wearing jumpsuits but no gloves or other exposure wear.

A rescue rope throw bag is tossed out to the canoe by someone, but it is neither secured nor used. There are so many rescuers in the canoe that there is little room for the victim, if he is rescued at all. Ultimately the man submerges and doesn't return to the surface. Since no one has marked his location either visually or on paper, the rescuers don't know where he is, especially after having broken up so much ice. The man's dog keeps nosing around one spot of the water, pointing to where his owner is. The

2

rescuers go to that spot. One of them, wearing a T-shirt, stands up in the canoe and decides on his own to jump in. It's amazing that he doesn't tip over the canoe and immerse his fellow firefighters. It's even more amazing that he doesn't drown himself.

It is only by much luck and providence that the man and the two girls are rescued without any serious rescuer injuries.

Two men in a small car on the ice plunge through, in an incident witnessed by snowplow operators, who immediately radio for help. The car's driver escapes through one window and begins walking across the ice. He falls through once, extricates himself, and then falls through again as rescue efforts begin. Rescuers are unable to reach him with ropes, because they can't get within a hundred feet of him across an open channel. A rescuer brings a ten-foot flat-bottomed boat to the scene and uses a claw-ended pole to break ice and move the boat forward.

By the time rescuers pull the man from the water, he has been immersed for forty-five minutes and has a core temperature of 82°F. During the helicopter flight to the hospital, he goes into cardiac arrest. Despite resuscitation efforts, the man dies. Although the passenger of the car is recovered by state police divers, it's possible that the driver could have been saved by ice rescue technicians with an ice board.

Working with water rescue teams for more than two decades and continually studying such incidents has given us the opportunity to observe the evolution and trend of the water rescue and recovery industry. This industry has made leaps and bounds over the past ten to fifteen years, and there has been a surge in water rescue and recovery training, equipment development, educational materials, and the number of departments that maintain surface and subsurface teams.

When we first started teaching surface ice rescue, it was a struggle to convince departments that members didn't belong on the ice without full exposure suits, harnesses, and tethering. Teams that actually had wet suits, ropes around their waists, and surface-ice protocols were state of the art. Lacking the tools and knowledge available today, we were teaching procedures that we now consider neither sufficiently safe nor effective. Having studied a myriad of incidents, it has become evident that a thorough understanding of different types of ice and ice formation is irrelevant to the rescue process since, if a call comes in, we already know that the ice is poor. We also know that the quality of the ice can change in minutes during any given rescue, both through natural causes and by our working on it. At the same time, the distance between the shore and the victim can include black ice, frazzle, open water, and snow. In training, therefore, we now only teach those procedures that work with all types of ice and open water, and we have replaced lectures on different types of ice so as to devote more attention to patient handling.

We have further stressed patient handling because it became obvious that water accident patients were typically being handled far too roughly, a result of rescuers operating out of their normal element. Now we provide all rescuers, fire, police, and EMS personnel with a strong foundation regarding cold-weather problems of physiology and patient management.

Today we teach teams and trainers that ice rescue transport devices and rescue flotation slings are as necessary for safe, rapid rescue as ice rescue suits, harnesses, and personal flotation devices. Now that rescuers are becoming better protected, it's time to accentuate innovative tools and procedures that will better protect the victim. Transportation devices and buoyancy slings allow us to reach, extricate, package, and transport a victim in a fraction of the time, and to handle him far more gently in the process. After all, aren't we doing all this to save someone?

When we tell teams and trainers that we expect the average rescue of a victim a hundred feet from shore to take less than ten minutes from the time of arrival, we're frequently told, "Impossible. It's ridiculous to expect that." Well, we thank all of our students of the past five or six years who have repeatedly shown us that, with the right procedures and equipment, such rescues are more than possible.

The techniques surrounding water rescue and recovery continue to evolve, as should training programs, standard operating guidelines, and the teams themselves. If you have been using the same tools and procedures for the past decade, it's time to learn more.

In the training milieu, we speak in generalities, since in all likelihood we have never seen you or your team. You may already be doing everything with maximum safety and effectiveness with the equipment and procedures available today. When we say "you" or "ice rescue team," we are referring to the average team and incident today. Your team may have equipment and procedures that are even more effective than those presented in this book. If you believe this to be true, we invite you to share your methods and information with ice teams worldwide. Only by sharing knowledge can our discipline be advanced.

4   Lastly, never take anything at face value. Never believe anything until you go out and try it yourself in a variety of ice and weather conditions. Before you purchase equipment or add a procedure to your standard operating guidelines, train and try it out on very weak ice, in open water, in wind, and in whatever other conditions your local area might present. Rescue is a hands-on activity, and it must be learned and tested that way.

1

No Ice Is Safe Ice

As the weather changes, a winter wonderland is nigh, bringing with it seasonal sports such as skiing, snowmobiling, sledding, ice fishing, hockey, figure skating, and simply the novelty of walking on frozen water. Every year, thousands of people experience near-tragedies on or near the ice, most of which go unreported. The ice cracks underneath them, and they suddenly find themselves immersed in frigid water. Miraculously, they pull themselves out, perhaps with the aid of a friend, and the matter is over in just a few seconds. Embarrassed, they find their way home, warm them-selves, dry their clothes, and thank their lucky stars. For the most part, that is how such incidents end.

Sadly, not all ice incidents end so simply. When public safety per-sonnel are dispatched to an ice incident, it has already been demonstrated that the ice is unsafe. The call never would have been issued if the ice were safe. If the victims can't get themselves out, then would-be rescuers are also in danger of crashing through and finding themselves suddenly immersed in frigid water.

Accordingly, you should always operate on the premise that no ice will hold. Treat any ice on which you must respond as if it will break anytime, anywhere.

We aren't going to take the time in the early stages of this book to discuss the viability of different types and thicknesses of ice. For the average responder, ice terminology means little when it comes to saving a life. It means even less when it comes to saving some-one three hundred feet from shore, where the ice density can

6

Always assume that the ice will break, be prepared to fall through, and then carry out the rescue as planned.

change every twenty feet. You should always assume that the ice will break, be prepared to fall through, and then carry out the rescue as planned.

Still, if you are responsible for determining whether or not a given expanse of ice is safe for public use, then you need a good working knowledge of its different types, as well as of the interaction between the weather and the density and thickness of the ice. Ice diving also requires a degree of specialized knowledge, since you must be able to decide whether the ice is safe enough for tenders to stand on or lie on, or whether an operation must be platform-based.

WHY DO WATER-RELATED RESCUES FALL APART?

Performing a rescue on ice is in many ways no different from other water-related rescues. Unfortunately, the following statements apply to far too many departments around the world:

1. We don't train for water rescue incidents.

2. We don't drill or practice for water rescue incidents.

3. We don't have the right gear for water rescue incidents.

4. We don't know what our particular job is regarding water rescue incidents.

5. We participate in relatively few actual water rescue incidents.

6. Our water rescue budgets are small or nonexistent.

7. We don't have SOPs or SOGs for water rescue incidents.

Imagine that a four-car, head-on collision has just occurred in your district and that you are on call. If you are with the fire department, you know exactly what to do even before you arrive at the scene. You know what personal protective equipment you will put on, what tools you'll need, and what necessary personnel and other resources are at your disposal. You know what your job will be when you get there. If you're an EMS responder, you know what to prepare on the way to the scene. You know you'll need backboards, collars, KEDs™, and other first aid equipment. If you're with the police department, you know that you'll

8   have to secure the scene, manage traffic, interview witnesses, and handle the paperwork.

When you get to the scene, you size up the situation, decide what needs to be done, and then go about doing it. Alerted by radio calls, even the personnel in the hospital's emergency room already have a good idea of what their job will be when the victim arrives by ambulance.

On the other hand, suppose a car with four passengers goes off a bridge and sinks into the icy water below. By the time you get to the scene, there is nothing to see, not even a ripple telling you where to look. What is the job of the fire department? What about the police? What should EMS personnel do? It's in these initial moments of confusion that the water rescue begins to fall apart. Faced with situations like this, responders lose all of their normal senses, for they can neither see, hear, touch, nor even smell the car. Most of your department's normal operating procedures won't apply. So what do you do?

There is only one job that absolutely must be done every time: You must be able to go home when the job has been completed. Stressing that point is the mission of this book. Your first priority is to keep rescuers safe.

A rescue is only as good as the safety procedures and training that precede it. Ice rescues can be risky, and there are no guarantees that rescuer safety won't be compromised even in a well-coordinated operation. Still, if you compromise your personal safety through negligence, shortcuts, ignorance, or bravado while attempting to save someone else, are you honestly doing a good job? Performing operations on ice and water without proper preparation often turns would-be heroes into statistics.

THE RESCUE BEAST

If a child just fell through the ice and you received the call over your scanner when you were only a few blocks from the incident, what would you do? Suppose that you're the first to arrive on the scene. The child is only a few hundred feet from shore. He is screaming and begging for help, and you have a hundred feet of rope in your truck. In your mind's eye, you're probably calculating a ready way to save this child.

Will the nature of the rescue beast take over, or will you use the procedures for which you've planned and trained? Will you tie the rope around yourself and a tree and then try to save him alone, or will you call for help and perform operational, shore-based procedures until more help arrives?

The rescue beast is present in anyone who is altruistic enough to want to save another's life. You wouldn't be in the rescue business if the safety and well-being of others weren't important to you. Still, where does your personal safety rank against your other needs and attributes? All too often, our hearts make us think of others first, yet so much can go wrong.

The rescue beast inside of you will try to convince you that your personal safety isn't of primary concern—getting out to the victim is all that matters. Unfortunately, the media feeds this mindset on a daily basis with countless stories referring to would-be rescuers as heroes, over and over again, regardless of whether or not those supposed heroes wind up dead; regardless of whether the actions of those supposed heroes endangered or ended the lives of the original victims; and regardless of whether those supposed heroes compromised the lives of other rescuers. Today's media caters to its audience in terms of sensationalism and hype, asking by inference, "Do you have what it takes to be a hero? Would you risk your life for another? Would you seek temporary fame by making a freelance rescue attempt?" Yet the implied message of the media seldom mentions that two victims are never better than one, and the rescue beast inside of you is no kin to your better judgment.

Clearly, if an unprepared rescue attempt fails, both the original victim and the rescuer lose. In the scenario described above, who is the first priority? The kid is screaming for help, but you cannot reach him with your tether line. What would you do? Would you wait until help arrives to extend your line, or would you disconnect and go for the rescue?

In purely logical terms, if the victim fell through the ice, what would most likely happen to you? If the victim can't get out of the hole, what makes you think that you could? Even if you can get to him and possibly get yourself out, what are the chances that you could get *both* of you out and back to shore? The answer is that, without training, equipment, and the proper support, *nothing* makes you any different from that victim.

Anytime you take a shortcut on safety, the danger exists that you will become a victim also in need of rescue. When one of our own falls into trouble, typically the whole rescue effort will turn toward him. The would-be rescuer thus becomes a liability to all concerned, particularly the original victim.

10   A hero isn't someone who botches a rescue and winds up dead at the bottom of a lake or a river. A hero goes home. A hero is someone who calls off a rescue attempt because it is simply too dangerous for the team's capability, or who waits until the total operation is ready so everything works the first time. Yes, the child is terribly important, but if you perform the operation right the first time, you only have to perform it once.

TRAINING MAKES THE DIFFERENCE

Having all the right equipment isn't enough. Training and certification are essential. After training, your team needs to drill and practice with the equipment in poor ice. There is no substitute for the reality of poor ice—ice that is crushing and crumbling all around you, making movement almost impossible. You must learn how the equipment works in different circumstances, feel the frustration of not being able to make the rescue go as planned, and appreciate the horrendous physical effort required in many ice rescue situations.

In regions where there is no ice available for practice, you can train in the grass or snow in an open area. Drag the gear, crawl across the ground, feel the physical effort involved. Train and practice on buoyed tarps set up over the water's surface.

Nothing replaces being trained professionally by certified, insured, experienced instructors. Good instructors understand how to train all kinds of students: those who learn by seeing, by doing, by hearing, or by taking notes. Good instructors need to foster a trainee's reflexive skills by creating rescue scenarios and situations that bring realism to the training program. Quality training includes repeated live-action, hands-on drills, with attention given to the reasons behind each procedure. Learning requires doing and understanding, not just watching and memorizing. Finally, in terms of organizations, training must take into account the different abilities, resources, budgets, needs, and responsibilities of each responding agency.

SUGGESTED SOGs

Can your lieutenant legally order you into a fully involved fire without turnout gear and SCBA? Can your chief legally order you

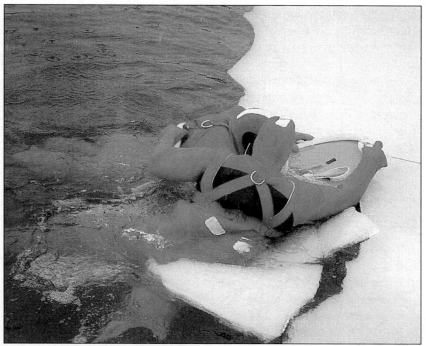

There is no substitute for training on poor ice.

into a haz mat incident if you aren't haz mat-trained and don't have the proper equipment? Can a police officer legally be fired for not jumping into the middle of a gunfight if he has neither gun, backup, nor protection? The answer to these questions is no, simply because public servants operate according to rules, bylaws, procedures, and guidelines.

By contrast, however, ask yourself whether a chief can order an unprepared firefighter onto the ice to save a child—something that happens almost daily in the wintertime. Why? Because there are few written rules, laws, procedures, or guidelines in place for water and ice rescue.

A department's surface ice rescue standard operating guidelines should state what your team can and should not do, the "should not do" parts being just as important, if not more so, than the "can do" parts. For example, a department's SOGs might specify certain conditions as determining a no-go response, such as moving water or drift ice; a partially or fully submerged vehicle; an ice canopy above a lowered water table; or an incident involving hazardous materials. Well-written SOGs also spell out the appropriate agencies to call to support an ice rescue operation, and they should specify when an operation changes from a rescue to a recovery. To reflect the true nature of these services, the terminology should be changed from "rescue" to "attempt to rescue."

Finally, the SOGs should also state that any rescuer, at any point, can say no to performing an operation because of insufficient training, a lack of available resources, extreme danger, or because he doesn't feel physically or mentally capable of performing the job safely. It is up to you, your peers, and your department to protect each other!

THE THERMAL HAMMER

Sudden, accidental immersion in cold water can affect the body and mind so rapidly and severely that it resembles being hit by a massive hammer. The thermal hammer can cause intense pain, great increases in blood pressure, uncontrolled gasping, confusion, disorientation, panic, and death. It pounds the victim's body, mental condition, and emotions. The thermal hammer can turn a strong, confident swimmer into a panicked, gasping, drowning victim in seconds. It can drown an adult in less than sixty seconds and a child in less than twenty. Improper handling of someone hit by this effect can result in needless injury and even death.

Thousands of small details come into play during water rescue. It's rarely just a matter of getting in there to get the victim. We need to understand the basics of what is happening to the victims and ourselves during accidental or planned immersion if we are to perform an ice or cold-water operation safely and effectively. If we do a good job of reaching a victim but then needlessly injure him during transport to the shore, we fail. If we don't understand what can happen to us during immersion, then we won't be as safe as we should be. Therefore, the cold stress, immersion, and patient-handling sections of this book should be read and studied by all public safety personnel who might respond to a cold-water or ice rescue or recovery, whether or not they have any medical training or responsibilities.

A typical would-be victim starts out by walking, skating, or playing on the ice. Suddenly, he hears the unmistakable sound of cracking ice. As if in slow motion, he realizes what is about to happen to him. Then, all too suddenly, he punctures the ice cap and finds himself immersed in frigid water. The shock of the cold

Thousands of details come into play during a water rescue.

water and the sheer mental trauma that he experiences will dominate the next few moments of his life.

If he was lucky, his hands and arms didn't go up as he went through, but rather, his arms remained low and helped him break his fall as he plunged into the water. If he wasn't that lucky, he could be experiencing the pain and shock of fractured ribs and other injuries in addition to the effects of the thermal hammer.

With a little more luck, he was moving slowly forward as well. If so, his forward momentum, combined with the immersion, wouldn't be enough to force him up under the ice roof, where his basic instinct could be his last. Why? Imagine yourself in the same situation. As you puncture the ice cap and become submerged, you

14   find yourself disoriented in the realm between light and darkness. You see dark and light spots under the ice. The ice distorts the light passing through it. The ice roof often provides a clear or bright-light effect, while the area where the puncture or hole is looks dark or even black. Light from the sun is reflected off the open water, while sunlight can pass through areas of ice, which then appear as light spots to the submerged eye.

Sadly, this effect is often considered a contributor to ice drowning, especially in situations where the victim is recovered at a distance farther from the hole than he should have been, given a slow current. As the victim submerges, he finds himself under the ice roof, and in his blurred vision, he sees light and darkness. He doesn't realize that going for the dark could mean life, and going for the light will most likely mean death.

Going back to the scenario, with a little more luck, the victim barely got his face wet, meaning that he didn't inhale a lot of water and close off his trachea. After a few attempts to get out, he realizes that it is a losing battle, so he wisely decides to hang on to the ice and wait for help to arrive.

Now it's your turn to act.

STUDY QUESTIONS

1. What sort of ice is safe?

2. As a rescuer, you should be prepared to fall through the ice and then _____.

3. Name some of the operational reasons that water-related rescues often don't go smoothly.

4. Imagine yourself to be the sole witness to an ice incident. Rather than acting rashly, it is best if you _____.

5. Without proper training, equipment, and _____, nothing makes you any different from a victim.

6. After training, is it most beneficial for your team to practice on good ice or poor ice? Why?

7. Name some of the reasons that a responder might legitimately refuse to perform an ice rescue operation.

8. Drownings often occur because, from underneath, the ice roof can look _____ while the hole looks _____.

2

Preplanning the Surface Ice Rescue Response

To perform safe ice rescues, the responding public agencies must be well prepared. This entails knowing what to prepare for; that is, preplanning for the potential incidents that they may face.

All responding agencies should develop response plans prior to actual emergencies. Because of the additional information that it provides, preplanning helps to create the safest, most effective SOGs possible. The process of preplanning brings the different responding agencies together, resulting in better cooperation and organization during actual incidents. Interagency preplanning is critical toward ensuring an effective incident management system, since the members of the various agencies must understand that there can only be one incident commander and one command post. Further, they must understand the organizational structure and hierarchies of the incident management system, as well as be able to communicate using its terminology and protocols. If such criteria are met, interagency cooperation and effectiveness will be facilitated.

The importance of interagency preplanning and follow-up drill sessions with debriefings cannot be overemphasized. Without effective and knowledgeable leadership, a simple ice rescue can become a disaster for all involved. Consider an incident involving a boy in the water, fifty feet from shore. How difficult is it to plan, command, manage, and perform such a rescue? With a well-trained and well-equipped team, the rescue should take less than ten minutes from the time of arrival on scene. As hundreds of students

18

A rescue should take less than ten minutes from the time of arrival on the scene.

demonstrate to us each winter, a rescue can be capably performed in less than that amount of time.

Unfortunately, not all teams are so well-prepared. Who shows up when an ice call occurs? Either no one or seemingly the entire world. Turf wars, battles for command, lack of equipment, lack of trained or knowledgeable personnel, difficulties in reaching the site, freelancing rescuers, uncontrolled media and bystanders, lack of interagency coordination, and a variety of other problems often arise. Inter- and intra-agency preplanning and drills can prevent most, if not all, of these problems. Without planning and drilling, at least some of these problems are almost guaranteed to occur, and then everyone loses.

WHERE AND HOW DO INCIDENTS OCCUR?

A major portion of preplanning ice rescue operations involves reviewing sites to which your agency has responded before. Several key questions always apply:

1. Where are the most common problem spots?

2. What are the best access and exit points?

3. Typically, how far from shore do the rescues take place in each area?

4. What have been the causes of ice incidents in the past?

5. What was the quality of the previous responses? What agencies responded, and what were the results? Were there too many responders or too few? Were the responders trained to follow the proper procedures, or was there chaos on the scene? Was the necessary equipment available? Were the rescuers put at unnecessary risk? Was there an effective debriefing afterward?

6. Did the IMS work, and if not, why? What needs to be done to prevent the same problems from reoccurring?

7. Are there any measures that the community can take to prevent ice incidents in the future, whether through education, law enforcement, or other means?

You may find that there have been no consistent patterns in the ice incidents in your district. On the other hand, you may discover that the same general problem has occurred year after year. For example, suppose you find that several incidents last year involved snowmobilers two miles from shore. The following actions might be mandated:

1. Long-distance ice operation training must be obtained.

2. A long-distance ice operation SOG must be written.

3. Long-distance ice operation equipment must be secured, such as the use of helicopters, hovercraft, or snowmobiles with sleds.

4. Rescuers must be further trained in patient handling, since it will take longer to transport patients to the ambulance and hospital.

5. Rescuers must be trained and physically fit for long-distance ice operations.

6. The county may want to pass laws to impose fines on snowmobilers who require a tremendous rescue effort, costing taxpayers thousands of dollars and damaging the environment with fuel spillage.

20   Has your department trained with its mutual aid departments, or have you at least collectively reviewed your options prior to the winter? If not, do not put it off! Also, whenever possible, train in areas where ice rescues have already been performed. Train throughout the year, especially if personnel turnover rates are high.

THE STEPS OF PREPLANNING

1. The first step in preplanning is to review past ice rescue incidents and find the potential ice problem areas in your district.

Determine where people most commonly congregate on the ice. Where do they skate and fish in the winter? Where might errant snowmobilers or skiers wind up, either by choice or by accident? Is there a particular location where vehicles have repeatedly gone into the water?

What areas in a given body of water are prone to having potentially weak ice? Where do geese and other waterfowl commonly congregate in the winter? Where are bubblers installed? Are the locations of any springs known? Are there any sewage or industrial pipes depositing chemicals or warm water that might weaken areas of the ice? Questions such as these will help you determine where the problem areas are in your district, for what sort of ice rescue operations your department should plan, and what preventive measures might be taken.

2. Once the potential sites have been located and marked on maps, determine the potential problems and hazards they might present.

An incident that involves a vehicle in the water requires a haz mat operation.

Ice rescues of snowmobilers may be compounded by trauma, as well as head and spine stabilization transport procedures. Ask yourself whether the site is difficult to reach because of steep embankments, trees, large rocks, fencing, ice, or deep snow. Is the problem area a long distance from shore? Is the body of water a long distance from the nearest road? Could a plane, bus, or train end up on the ice? How about a truck with hazardous materials?

3. Next, determine what is required to make ice operations viable.

What are some of the compounding problems that rescuers might find, and what is needed to deal with them? Vehicles, victim trauma, weather, and the location of the incident can all present a myriad of problems to the rescuer. Some of the more common complicating factors are described below.

4. *Make all personnel and equipment ready to do the job.*

What time of day are calls most likely to occur? If most of the rescuers are volunteers, how many can be available during those times, and where will they be coming from? Do they keep warm clothing, boots, gloves, and hats in their vehicles at all times?

Are all personnel trained and certified at minimum to the Awareness level of ice rescue? Are sufficient numbers of the right rescuers trained to the Operational and Technician levels, and will they be the ones who will be available during high-risk times? The term "right" rescuers means right for the job—those who are physically fit, trained, and capable. Personnel who are large, heavy, unfit, older, hypertensive, insulin-dependent, poor swimmers, or asthmatic do not belong out on the ice performing ice rescues.

Are a sufficient number of EMTs and paramedics trained to the Awareness level?

Do EMS and hospital emergency department staff have written protocols for long-term drowning, near drowning, immersion hypothermia, cold stress, fuel contamination, and other related problems? Do the responders all know what their responsibilities and duties are, and have they practiced them? Are the police trained to recognize abuse, neglect, and homicide by drowning? Are fire department officers sufficiently trained to manage ice incidents?

COMPLICATING FACTORS

As mentioned above, any number of factors may complicate an ice rescue operation. Although the list of potential variables is virtually limitless, the following are provided as being among the more common.

Vehicles

To be fully competent, rescuers must have had sufficient training to deal with the specific hazards of vehicles in the water, such as rescuer entanglement and entrapment; objects containing air that act as projectiles; jagged metal; fuels; and patients with trauma or entrapment problems. The responders must be fully aware at all times that if they are close to a submerging vehicle, they can easily be pulled down with it. Similarly, they must remember that aspiration of fuels can result in lipoid pneumonia and other health

22

problems. Without proper training, rescuers may attempt to use tow cables attached to a car, which could break if the vehicle submerges, thereby causing serious rescuer injuries. The dangers of immersed and submerged vehicles are often hidden behind ignorance, yet the potential consequences of seemingly routine procedures are very real.

Similarly, ice shacks with heaters present a number of problems. An ice shack can puncture the ice with its occupants still inside, and the resulting operation may be one of extrication, as in the case of a vehicle. People of all ages and sizes go ice fishing, so the problems that occur with them can include all types of medical emergencies, as well as hypothermia.

Trauma

Not all ice calls involve immersion. Departments may need to handle everything from fish hooks in eyes to broken bones, burns, frostbite, and major trauma. Of course, any one of these problems could be compounded by immersion.

If trauma may be involved, perhaps the best choice of personnel would be those first responders or EMTs who truly understand the importance of gentle handling, maintaining open airways, the cessation of major bleeding, and other life-threatening matters. You should determine whether any cross-certified EMT/ice rescue technicians are available. If not, perhaps your department or district needs to train some. Another consideration is the availability of ice transport devices that would be suitable for trauma or head- and neck-injury patients.

Similarly, have Operational- and Technician-level rescuers planned and practiced procedures for handling hypothermic trauma patients on and off the ice? Have they done this wearing ice exposure suits and gloves? What problems could victims wearing snowmobile suits, helmets, or skis present?

Alcohol

During your fact-finding, you may very well discover that alcohol has been the largest common denominator in all past ice incidents in your district. If so, rescuers may need to be prepared to handle less-than-helpful victims, perhaps even belligerent ones. Intoxicated persons are more apt to take risks. Such behavioral aberrations can be especially difficult in ice rescue situations. Physiologically, intoxicated victims face a greater risk of drowning, since their trachea is less likely to close (laryngospasm) when they aspirate water. Also, they may vomit, causing additional airway problems. Intoxicated victims may not understand the severity of their situation and may not be able to follow even the simplest commands of their rescuers. More than likely, their reflexes, judgment, coordination, and senses are impaired.

Weather

If a call comes through during a blizzard, special strategies and tactics will be required. If snow and wind prohibit tenders from seeing or hearing technicians at the end of the tether line, for example, then line-pull signals will likely be the prime means of communication. Perhaps additional wind-blocking hardware will be needed. If the responders and victims are being pounded by hail, then helmets and other protection may be required during the rescue and transport to shore.

What is the strategy if the roads are closed? Snowmobilers could still end up requiring help on an ice-covered lake when blizzards make road conditions treacherous for ambulances and rescue vehicles. It's possible that a large number of volunteer rescue personnel will be tied up clearing roads as county snowplow operators. If so, what happens to the immediate availability of surface ice technicians if a call occurs?

Deep snow not only weakens ice, it also hides where the ice starts. You should preplan how the hot zone and warm zone can be recognizably marked once the shoreline has been found.

If severe winds are known to occur over a given body of water, what special equipment and SOGs are in place to deal with them? If you have ever watched a 260-pound man blown across thirty feet of ice into open water, you'll appreciate the importance of this consideration, as well as tools such as ice cleats, ice awls, and spiked ice poles.

Ensure that there is enough personal protective equipment to protect the rescuers from the chilling effects of rain, wind, and snow. As will be discussed in the medical sections, cold rescuers are far more apt to make poor decisions and take risks. If it is a large, complex operation, obtain propane heaters for the rehab area. Make hot soup and warm, noncaffeinated drinks available to help keep personnel warm.

Any operation will change after nightfall. Ensure that there is sufficient lighting for the staging areas, the command post, and the different sectors of the operation, and have a plan for any large-area surface ice search that is to be made during the dark hours.

24

Snow not only weakens ice, it also hides where the ice starts.

If possible, deploy a helicopter with a spotlight. The ice rescue technicians should have waterproof hood lights mounted on their suits to allow them full use of both hands.

As with any other aspect of these types of services, your department should have established policies for water-related incidents during electrical storms.

Access Points and Staging Areas

In terms of shore access, decide on planned shore access points, and have the necessary tools to make those locations usable. Find the shortest route to the middle of each body of water in your district. To ensure that all personnel are aware of these preferred routes and staging locations, draw maps and label the sites. You should drill at these sites so as to familiarize personnel as to the best place to stage and park vehicles while doing as little property damage as possible and keeping personnel safe.

If a body of water is completely surrounded by thick forest, you may need chain saws to clear a staging area. If deep snow or icy shore areas are a potential problem, be prepared to handle them as well.

Ambulances may need chains on their tires. Perhaps only four-wheel-drive vehicles can be used. If helicopter transport is necessary, where can a landing zone be created in the winter? When the terrain leading to the shore is steep, rescuers should be properly trained in belaying and tethering techniques.

Remote Locations

Snowmobilers, hunters, and cross-country skiers may find their way to ice-covered lakes and ponds deep in the woods, miles away from any road. Rescue attempts in such locations are compounded with difficulties. First, witnesses may have to travel miles to reach a phone, given that they don't have immediate access to cellular phones or radios.

When the call is received by Dispatch, the immediate question becomes, How will rescuers reach the site? Do they even know how to get there? What is the best route? Are snowmobiles with sleds accessible to transport equipment and personnel?

The first step is to know about all these bodies of water and make sure that they are clearly documented on maps. In the event that a child or adult is reported missing in the winter, any number of adjoining areas may need to be searched.

Next, determine the best routes to these sites in the winter, as well as the type of transportation needed to reach them. Consider all of the viable options and test them. If no public safety agency has snowmobiles, for example, see whether other modes of transport are available in the event of an emergency. Horses may be the only viable option. Perhaps local snowmobile dealers or ranches would be willing to help by providing transportation during drills and actual incidents. Make sure rescue personnel are competent enough to handle any special equipment or animals that they are offered.

It's possible that another mutual aid department has an easier or shorter route to the location. Perhaps another department, team, or agency has better transportation devices. If such is the case, work with your department's mutual aid agencies to create the most effective and viable SOGs, then train and drill together.

Once appropriate routes and transport devices have been found, figure out how many personnel are needed and what equipment must be brought to the scene. Determine how to load these resources on the transport devices, then go for a test run. Note how long it takes to load up the gear and personnel, plus the travel time to the destination.

Lastly, you must devise and test a plan for transporting the victims from the remote site. If a helicopter cannot land near the site, you must have several plans for transporting the victims to the nearest road. If only one victim is involved, perhaps an extra transport device or a sled can be brought in. If not, be prepared to leave personnel and gear behind to get the patient to medical help as quickly as possible. If there are multiple victims, personnel and gear will likely have to be temporarily left behind. Make sure that any such personnel are prepared to withstand the cold environment for whatever length of time they'll be exposed to it prior to being retrieved.

Ensure that the transport device is adequate for a patient with a head or back injury, trauma, hypothermia, or other problems. Have sufficient dry thermal protection for the ride back for both rescuers and patients. At least one rescuer with EMT or higher status should be on the scene in any event.

EQUIPMENT

Toward the end of fall, it is time to outfit gear and rescue vehicles for ice operations and to make sure that everyone knows where everything is. Make laminated cue cards for communication signals between tenders and ice rescuers on the ice, and mount them on PFDs with lanyards. Practice these signals.

If you have sized ice rescue suits, ensure that the sizes are clearly written on both the suit and the suit bag. Have profile slates and other documentation paperwork ready and easily accessible. Slates should have writing instruments. Make sure that any rope bags are clearly marked for length, and that the line itself is marked in increments and is in good working condition. Personnel should know which bags to use and how to read the line markings.

Check on equipment log sheets to be sure that all equipment is functional and in its proper place. Set up the gear for easy and rapid access, with each complete tender-rescuer's gear stored and set up in individual bags that can quickly be pulled off the truck and used. A common mistake is to store the same type of gear together. Doing so means that, for a tender and rescuer to dress, they must go to the PFD area to grab a PFD, then go to the harness box to get a harness, then reach for a suit in the suit locker, and scramble for all of the other necessary gear as well. Such scrambling about wastes time and increases the chances that gear will be lost and forgotten.

Finally, make sure that all of your department's equipment meets the requirements of the standing SOGs.

28   STUDY QUESTIONS

1. Name some of the key questions that always apply when reviewing ice incident sites for preplanning purposes.

2. What are the four steps of preplanning?

3. According to the text, among other health problems, the aspiration of fuels can result in _____.

4. Ideally, your team should include ice rescue technicians who are also cross-certified as _____.

5. Typically, the single largest complicating factor in ice incidents is _____.

6. An intoxicated victim faces a greater risk of drowning because _____.

7. When determining shore access points, you should find the shortest distance to _____ of each body of water in your district.

8. When the terrain is such that personnel can only reach the victim on foot, rescuers should be properly trained in _____.

9. Name four essential considerations when preplanning ice rescue operations in remote areas.

10. You must ensure that whatever equipment you're using conforms to your department's _____.

3

The Three Levels of Training

Multiple agencies respond to ice rescues, yet their departmental standards are often dissimilar. This chapter includes information on the Awareness, Operational, and Technician levels as described by the National Fire Protection Association (NFPA), which sets standards of performance capability and training. The immediate duties of each level are listed below and discussed in detail in later chapters.

AWARENESS LEVEL

Anyone who might respond to the scene of an ice rescue should be trained to the Awareness level, which involves a minimum of four hours of training. The Awareness level is the lowest level of training, the main goal of which is to prevent further injuries and problems from occurring at the scene. Public safety personnel at this level should be equipped with appropriate personal protective equipment. Agencies whose personnel are most likely to be trained to the Awareness level include police, EMS personnel, park rangers, and some fire personnel.

Proper training and protective equipment are especially important for ice rescue incidents, since would-be rescuers who attempt to go out on the ice without more advanced training and equipment are likely to end up in the same situation as the original victim. Awareness training will help prevent rescuers from

30 becoming victims and thus hindering the operation.

The Awareness level involves the ability to perform and delegate the following:

1. Know where to go to access the scene and stage the operation, based on predetermined, documented staging areas and the location of the incident.

2. Identify and use personal protective equipment.

3. Perform a rapid scene assessment to identify the hazards to life, determine who needs to be dispatched to the scene, and decide what kind of response is necessary. A rescue technician trained to the Awareness level should be able to (a) identify hazards to life and health; (b) assess the number of victims and, if possible, their conditions; (c) know where and how to get the needed assistance; and (d) mark a spot on the shore in front of each victim.

4. An Awareness-level rescue technician should also be able to implement the incident management system (IMS), including establishing a command post and incident commander, as well as staging areas, staging officers, safety officers, and other officers and modules as required by the size and complexity of the incident.

5. Call for the appropriate personnel and response, including a dive team.

6. Secure and manage the scene to maintain scene safety. This entails determining and marking operational zones. The incident scene can be divided into three zones: hot, warm, and cold. The hot zone is on the shore, within ten feet of the water's edge, and may be entered only by those with Technician-level training. The warm zone is where the operational efforts are staged, and the cold zone is the shore area behind the warm zone. The cold zone is established for the media, family members of the victim, and bystanders. The boundary between the cold and warm zones should be marked and, if possible, taped off. If deep snow makes it difficult to see the line dividing the hot and warm zones, mark that boundary as well to prevent anyone from inadvertently stepping out onto the ice.

7. Keep the media, family members, and bystanders within the cold zone.

8. Secure and check the condition of any witnesses.

9. Determine and direct the positioning of incoming vehicles to form windbreaks, as well as to prevent blocking the exit routes of ambulances.

10. Reassess the victims, if possible. If submergence has occurred, determine whether the situation is one of rescue or recovery, based on the time of submergence and your SOGs.

11. Commence communication with the victim if Operational- or Technician-level rescuers haven't yet arrived.

12. Interview witnesses and document the information on a profile map if Operational- and Technician-level rescuers haven't yet arrived.

Since ice rescues aren't a regular occurrence, make a checklist for Awareness-certified members to keep in their vehicle or wallet during the ice season. Such a checklist can be put on laminated ID-size cards. It may also be necessary to keep the map of possible staging areas in their vehicle, depending on the size of their jurisdiction.

OPERATIONAL LEVEL

Rescuers trained to the Operational level possess additional knowledge about the specifics of ice rescue, and they are able to perform shore-based rescue procedures. Their duties are to:

1. Finish implementing the IMS, with an ice rescue technician and a diver as advisors at the command post, if possible.

2. Further assess conditions and hazards.

3. Create a profile map of the incident and the location of the victims, if not already done.

4. Interview witnesses and record information on the profile map, if not already done.

5. Identify available resources and ensure adequate response.

6. Identify and use necessary personal protective and shore-based rescue equipment.

7. Develop a shore-based plan of response with available resources per the SOGs.

8. Attempt to establish communication with the victims.

9. Further assess the status of the victims.

10. Implement the planned shore-based response—help the victim perform self-rescue procedures.

11. Assist Technician-level rescuers with shore operations and tending.

TECHNICIAN LEVEL

Rescuers trained to the Technician level have direct exposure to the hazards of ice and water. The moment it becomes necessary to set foot on the ice, the operation requires personnel trained to this level. Whenever a rescuer steps onto the ice, the entire operation changes, since the rescuer is now exposed to the same hazard that befell the victim. Ice rescue exposure suits, lines, harnesses, trained tenders, backup personnel, and more will now be required.

Whenever a rescuer steps onto the ice, he is exposed to the same hazard that befell the victim.

The duties of surface ice rescue Technicians are to:

1. Perform a risk-to-benefit analysis of the incident so as to make a go/no-go decision.

2. Perform self-rescue procedures. Technicians must be able to swim and handle the exertion. They should, therefore, have an annual physical examination to check for hypertension, asthma, and other conditions that might pose limitations.

3. Advise and coordinate operations with the incident commander.

4. Coordinate operations with the responding dive team.

5. Develop a plan of response, given the available resources and per the SOGs.

6. Implement the planned response.

34 **STUDY QUESTIONS**

1. What are the three levels of ice rescue training?

2. Police, EMS personnel, and park rangers are most likely to be trained to what level in terms of ice rescue?

3. The hot zone commences _____.

4. The hot zone may only be entered by personnel at what level of training?

5. True or false: An Awareness-level responder, if first on the scene, is authorized to institute the incident management system.

6. Operational efforts are carried out within what zone?

7. The cold zone is established for _____.

8. Line tenders must be trained at least to what level?

4

Who Responds to Ice Rescue Incidents?

POLICE

The police are often the first ones on the scene, since they are normally mobile within their territories. At minimum, they should be trained to the Awareness level, although it would be helpful if at least some officers were trained to the Operational level as well. Park rangers can also be considered to be within this category of responders.

FIRE DEPARTMENT

Because providing rescue services is part of its normal role, the fire department is normally the second entity to arrive at the scene of an ice rescue. Whether volunteer or municipal, a department has to mobilize, and even in the best conditions, this requires a bit of time. Firefighters should be trained at least to the Operational level, and they are the most likely personnel of any agency to be certified to the Technician level. Responding fire vehicles should carry the necessary ice rescue equipment on board.

EMS PERSONNEL

If they have been adequately dispatched, medical units usually arrive at the scene of an ice rescue close to the time that fire

35

companies do. EMS personnel should be trained at least to the Awareness level, and they should have additional training in the management of cold stress, immersion hypothermia, near-drowning, drowning, long-term drowning, and scuba-related injuries. They should also be trained in how to take care of rescuers, such as knowing to check the blood pressures and condition of the technicians and divers before they suit up as well as when the job is over.

THE DIVE TEAM

The dive team should be called for any operation, even a surface rescue, since divers may be needed to help keep the victim buoyant or in case he submerges. On the scene, the team should:

1. Use standard ice rescue diving procedures.

2. Obtain a briefing from the incident commander and prepare for possible diving operations.

3. Stage wherever Command dictates to avoid hindering the surface operation.

4. Evaluate the area. Can the victim be reached from the surface, or will the operation require taking a route below the ice?

5. Discuss options with the incident commander.

6. Determine whether the dive tenders need to be on the ice or in the water so as to perform their duties properly. Are they dressed appropriately for possible immersion? Is a boat platform required?

7. Have EMS personnel check out the divers. Have tenders dress the divers and keep them sheltered. Keep dive gear off of the snow and ice to reduce the chances of freeze-ups and items being lost. Secure all safety lines, and double-check all dive gear and personnel.

The dive team should have trained with surface rescue personnel prior to responding to an actual emergency. There is no substitute for joint training, since every second counts during real operations. A lack of joint training may result in an ineffective IMS, poor communication, confusion, chaos, and tragedy.

If the ice is very poor, deploy a diver under the ice to keep the victim afloat until the surface rescuers can reach him.

Scuba Diver Deployment for Surface Rescues

Fully dressed, properly trained scuba divers don't belong on or under the ice until the incident commander gives the word and the planned response is known. Will they perform a subsurface recovery of a drowned victim, or are they required to conduct a subsurface operation in support of a surface rescue? The ice may be so poor that there is no access across it without possibly losing the victim, and rescuers may be falling through it with every step. Hence, the divers must use the most direct access when the ice is falling apart, deploying just under the ice, surfacing by the victim, and providing him buoyancy. Once this is done, the surface rescuers can approach from any direction, by any means. Even if they destroy the ice, the diver still controls the victim with buoyancy at the surface.

Establishing buoyancy on contact with the victim is critical. If the surface rescuers cannot reach the victim because of poor ice, a diver can be deployed just under the ice roof to hold the victim up and keep him afloat, thus buying time for the surface crew to reach him.

Always remember that ice diving certification is not the same as surface ice rescue certification. Dive teams should not consider themselves competent and capable of surface ice rescue just because they are ice divers. Dive teams should obtain surface ice rescue Technician training and certification to be fully competent.

STUDY QUESTIONS

1. Of all responding agencies, _____ are normally the first to arrive on the scene of an ice rescue incident.

2. _____ are the most likely personnel of any agency to be trained to the Technician level.

3. EMS personnel who respond to ice rescue incidents should have additional training in the management of what cold-related conditions? (Name several.)

4. True or false: The dive team should be called to any ice rescue operation, whether the ice has broken or not.

5. No diver should deploy on or under the ice until _____.

6. When the ice is poor, the primary reason for deploying a diver is to _____.

7. To be fully competent in ice rescue operations, members of dive teams should obtain _____.

5

Managing an Operation

Incident management system organization and terminology may differ not only between departments but also within a given department as well, depending on the type of incident at hand. The following are some ideas and examples of an IMS that may prove helpful to your own ice incident responses. Throughout, the incident commander is referred to as the IC, or simply Command, as he would normally be referred to over the radio. The IMS might be called the incident command system (ICS) or the incident command management system by your department, but these are just different terms for essentially the same thing.

This chapter isn't designed to teach you how to plan and implement an IMS. Rather, its goal is to present specific concerns during ice incidents. We have been told of training programs that spend six or seven hours on IMS, yet when the next water-related incident occurs, the system is barely implemented and quickly falls apart. IMS training must go well beyond the classroom, with live-action, hands-on incident scenarios involving multiple agencies. An IMS isn't worth the paper it's printed on if it isn't tested and practiced in a realistic way.

Without effective leadership, all the best equipment and trained personnel in the world won't guarantee a safe and successful operation. Since ice rescue incidents and drills are typically rare occurrences, solid leadership is critical. A well-designed and properly implemented IMS, led by a skilled and knowledgeable incident commander, can prevent rescuer casualties, can ensure

40 the quickest time for rescuers to reach the victim, and can ensure that the victim is handled properly during transport to the ambulance. Without an effective IMS, unprepared rescuers could end up on the ice, the victim's location could be lost, and the victim could be further injured by mishandling.

In every case history that we know of in which rescuers were killed or injured during an ice rescue, an effective IMS with ice rescue expertise was obviously lacking. Presumably, an effective IMS led by commanders with the necessary expertise could have prevented those casualties from occurring. Surface ice rescue is actually one of the safer types of rescues, as long as it is performed correctly with the proper equipment, procedures, and trained personnel. It is up to the IMS to ensure that these three factors are present and will combine for a smooth operation. If any one of them isn't fully ready and in place, a good IC will shut down the process until the situation has been remedied. If a given emergency is beyond the available resources, then the operation should be delayed until the appropriate agencies can respond. It's up to Command to make this happen. The IMS should also support any rescuer who doesn't want to go out on the ice for any reason.

THE INCIDENT COMMANDER

There must be only one Command and one command post. A unified command is necessary to coordinate the efforts of all of the agencies involved. This unification is especially important for water-related incidents, where the duties and functions of different agencies are often less than clear. When responding agencies are ill-prepared for water-related incidents, anyone from bystanders to law enforcement officers may be in the water or on the ice, attempting to reach the victim. Knowing who does what, where, when, why, and how is crucial to smooth rescue operations. Perhaps a police dive team and a surface ice rescue fire company simultaneously arrive on the scene. Which agency should undertake the mission? Or, more appropriately, how can they work together for optimal results? Turf and jurisdiction wars shouldn't be resolved during actual emergencies; they should have been decided during preplanning sessions. Moreover, an effective IC and IMS will be able to prevent such frictions from encroaching on an emergency scene. One county in New Hampshire, for example, implemented the rule that, as long as an

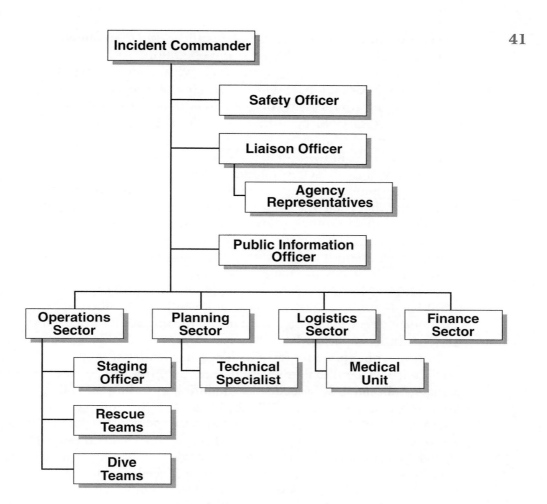

A typical incident command structure for an ice rescue operation.

incident is in the rescue mode (which includes up to one hour of submergence without it being a known crime scene), then the fire department has jurisdiction. Once the incident shifts into the recovery mode, then the police are in charge.

Command must be able to step back and observe the entire operation to devise the plan and ensure that it is carried out according to the department's SOGs. He must delegate jobs effectively and not fall into the common trap of getting involved with hands-on activities. Further, he must be able to ask the right questions and listen to the advice of the ice rescue technician representative, the medical unit leader, the dive team representative, the witness interviewer, and the safety officer, and then integrate this information into a viable plan and operation.

Command should receive notes from witness interviews and profile map information, and he should commit the plan of operation to paper. An incident command board helps to ensure that all of the necessary officer positions are delegated, assignments are given out, and resources are obtained.

An effective IC knows not to rush the operation, and he knows how to slow down rescuers and officers who do. The manner of all of the respondents should be professional, meaning that every movement that each one makes should have a purpose. Rushing typically results in mistakes and broken equipment and usually ends up wasting precious time. Rescuers must by creed perform the right action, in the right way, in the right amount of time. It is up to the IC to ensure that that happens.

The first-arriving officer usually assumes the role of Command until passing the position off to a senior officer, an officer with greater ice rescue expertise, or to a peer if the original IC is needed in some other capacity.

In terms of experience, an incident commander should have surface ice incident expertise. By nature, he must also be observant and should possess leadership qualities, including the ability to delegate duties to others. Whatever the IC does not delegate is his own responsibility. Leadership abilities are essential for the first job of the IC, which is to confirm that command has been established. He must formally announce that command has been taken and by whom. A common problem experienced during water-related incidents is that few personnel know who is in charge, which too often results in multiple failed freelance rescue attempts.

Command, the command staff, and other officers must be easily recognizable, which may be difficult when they are bundled under hats, scarves, and mufflers or are covered in snow if conditions are bad. Color-coded IMS vests can help alleviate this problem significantly, especially if the position of IC is passed off several times. In the winter, it is a good idea to keep extra hats, gloves, and other necessary personal protective equipment, plus command slates and boards, where the vests are stored.

If the IC is replaced for any reason, it is up to the new IC to decide whether the operation is going as it should. The new IC should not be afraid to change the operation if it is unsound. Prolonging an unsafe or unsound rescue procedure will only make it more difficult to correct. In making such decisions, he must evaluate whether there are enough, too few, or too many resources on the scene, as well as whether the right resources are available for any alternate operation.

On arrival at the scene, an incident commander should:

1. Obtain a briefing from the previous IC or the first responder to find out what occurred, what is occurring, what resources have been committed, what the hazards are, what needs to occur, and what resources are en route.

2. Assess the status of the incident and establish objectives and probable resource requirements.

3. Give an initial briefing, set up an IMS structure, and make officer and duty assignments.

4. Authorize the plan of action and make a go/no-go decision based on a risk-to-benefit analysis.

5. Communicate with and manage the command staff.

6. Approve requests for and the use of resources.

7. Approve the release of information to the media by the liaison officer.

8. Approve plans for demobilization.

The next step is to establish a command post that is easily accessible, visually obvious to all involved, and usable in adverse weather conditions. The post must be staffed at all times.

Documentation on the status of resources and officers, as well as a visual master map of all the different sectors of the operation, should be available to the IC in the command post. Any new information should be recorded immediately; otherwise, the operation could quickly become overwhelming. Never try to play catch-up. Situations can change rapidly: Suits might leak, thereby forcing technicians out of service; additional victims might be discovered; a surface operation might become a subsurface operation in seconds; a needed incoming company might be delayed somewhere; or the weather could change for the worse. Documentation makes it easier to change strategies and tactics. The command post receives, correlates, transcribes, and responds to all of this important information.

A rescuer at the Technician level and a member of the responding dive team should be at the command post to provide the necessary information to the IC so that he can make knowledgeable decisions. The command post should be staged in a location that overlooks the entire operation, is easily seen, and is accessible.

Although the incident itself may not require many members, it may take extra officers and supervisors to manage the number of personnel and agencies that respond. It is best to have one leader, or supervisor, for every three to seven members. The concept of manageable span of control is especially important for water-related incidents since, as discussed earlier, such operations are nonroutine, the equipment is infrequently used, and few personnel may be sufficiently familiar with the pertinent SOGs.

For large or difficult surface ice incidents, the IC personally assigns three officers to the command staff: a safety officer, an information officer, and a liaison officer. By fulfilling their respective functions, these officers help prevent the IC from becoming overburdened.

SAFETY OFFICER

The safety officer identifies, assesses, and isolates hazardous locations and conditions. This officer is responsible for the safety of all personnel on the scene and has the authority to prevent or stop unsafe acts if they are imminently hazardous.

The duties of the safety officer are to:

1. Obtain briefings from the IC.

2. Assess the safety of the operation.

3. Participate in planning at the command post.

4. Review the plans for the operation and identify any proposed actions that may prove unsafe.

5. Investigate any on-site accidents that occur during the incident.

6. Document and log all information relevant to the role of safety officer.

Since no one should be in the warm zone without authorization, PFDs, and other necessary personal protective equipment, the safety officer should post someone at the entrance to log personnel as they go in and out. This procedure is especially important at night, during storms, or any other time that someone might be able to go out on the ice without anyone else knowing about it.

At the entrance to the hot zone, the safety officer should post a checker—someone to verify that the ice rescue technicians are properly dressed and equipped. Each rescuer should be wearing a fully zipped ice rescue exposure suit, a tethered harness, a rescue flotation sling, and other such equipment. The ice rescue technicians should have their blood pressure checked before dressing, and each should have a chief tender to help him dress. The chief tenders must stay with their technician counterparts until they are dressed down and checked by EMS personnel.

The safety officer should ensure that the ends of all tether lines are secured to immobile objects and that there are enough properly equipped line tenders on each tether, all wearing gloves, PFDs, and other necessary equipment.

It is also the duty of the safety officer to ensure that ambulance personnel are waiting inside their vehicles rather than standing around without PFDs, getting cold during the rescue effort. They should have the appropriate transport equipment ready to take the victim from the shoreline to the ambulance. Witnesses may also need medical attention. If medical personnel might approach the shoreline, they need to be wearing the appropriate personal protection as well.

The access routes to the shoreline and the ambulance must be free of obstructions and hazards. Otherwise, the safety officer must either change the routes or delegate personnel to remove the problem. If possible, he should make sure that holes left in the ice after the victim has been recovered are then marked or staked off so that they don't later pose a threat to the unwary.

Like the IC, the safety officer must step back and continually observe the big picture of the operation. In short, he is responsible for enforcing all SOGs pertaining to safety. Strong observation skills and a thorough, working knowledge of the department's ice rescue SOGs are prerequisites for the task.

INFORMATION OFFICER

The IC and the operation itself must not be hindered by the media, bystanders, or the victim's family members. The information officer serves as a buffer and mediator between these groups. The information officer does what is necessary to keep media and other people not directly involved in the operation in the cold zone, a duty that may require the assistance of law enforcement personnel.

46   The basic responsibilities of the information officer are to:

1. Obtain briefings from the incident commander.

2. Assess the incident and document the information summary.

3. Obtain IC approval to release information.

4. Arrange meetings between the IC and the media.

5. Keep the media and civilian bystanders in the cold zone.

6. Fill out the appropriate documentation.

7. Provide appropriate information to the victim's family members at the scene.

LIAISON OFFICER

Effective coordination between all responding agencies at the scene of an incident is imperative. The liaison officer serves as the link between the IC and other agencies. He acts as a diplomat between agencies that aren't accustomed to working within the IMS. He also determines where each agency can stage, ensures two-way communications, and prevents other command posts from being established. The liaison officer's basic responsibilities are to:

1. Obtain briefings from the IC.

2. Locate each agency's representative and establish two-way communication so as to foster a coordinated effort.

3. Obtain additional outside resources as requested by the IC and fulfill agency needs as requested by agency representatives.

4. Monitor the operation to prevent, discover, and remedy any interagency problems.

5. Fill out the appropriate documentation.

STAGING OFFICER

Staging areas are places where personnel and resources are stationed until they are given assignments; where personnel ready themselves and their equipment once an assignment has been given; and where personnel dress down and repack their equipment after the incident. There are many advantages to having a staging area, particularly for water-related incidents. A staging area provides for easy accounting of personnel and equipment while guarding against freelancing and the premature, improper deployment of resources. By staging apparatus in controlled locations, the threat of equipment loss or damage is also decreased.

Each agency might have its own staging area. In a water rescue scenario, the fire department, police, EMS, divers, haz mat, and service groups might all be staged separately.

Each staging area is controlled by a staging officer, whose responsibilities are to:

1. Obtain briefings from the IC or the liaison officer.

2. Go to the designated staging area and set it up.

3. Establish a check-in and check-out procedure for personnel.

4. Provide resources when requested by the IC.

Although the IC controls the entire operation, each agency controls its own personnel and resources. For example, the IC may present the following: "A nine-year-old boy punctured the ice at 1730 hours and is now immersed and hanging on to supportive ice approximately sixty-five feet from shore (Sector 1). The boy's thirty-four-year-old father punctured the ice attempting to save him and is now submerged approximately thirty feet from shore (Sector 2). Deploy a primary and a backup ice rescue technician to the boy in Sector 1, with two shore-based tenders on each rescuer's tether line. Deploy the primary technician with the ice rescue board, which will be used to extricate the boy and transport him directly to the ambulance. The dive team will deploy to Sector 2. Dive tenders will don surface ice exposure suits and harnesses, and they will tend from the ice with tether lines secured to ice pitons. The backup tender will serve as profiler. The primary and backup divers will crawl to the hole with their gear on, once they are tethered and the tether lines are secured by the tenders and pitons. The ninety-percent-ready diver will sit on a piece of carpet ten feet from the hole with his gear fully ready, next to him on the carpet. He will be tethered with a harness and a line that is secured to a piton. An inflatable boat will be ready to transport the victim when found. Once the diver has given the found-object signal, the boat will be pulled to the hole with a pulley secured to the ice piton and three boat tenders. Gently roll the victim into the boat horizontally and carefully place him on the backboard.

Ambulance personnel will take possession of the victim at the warm/cold zone boundary."

Each staging officer then assigns members of his team to fulfill each task. The surface ice technician team officer assigns two chief tenders, two line tenders, and two rescuers to dress, and he assigns a member to tether the ice board and ensure that it is rigged properly. The dive team staging officer assigns divers and tenders to ready themselves for their tasks. The company with the inflatable boat assigns members to bring it to the shoreline, rig it with a tether line, and to coordinate with the dive team to set the piton and pulley. The medical unit leader sends EMTs to the dive and technician staging areas to take the blood pressures of the rescuers before they dress. He also sets up the rehab station, briefs the duty crew that the victims will be brought directly to each ambulance by fire personnel on an ice rescue board and a backboard, and assigns an EMT to ready the oxygen humidifier in an ambulance with heat packs once they are informed that a victim has been recovered.

AGENCY REPRESENTATIVE

A representative from each responding agency should check in with the IC or liaison officer at the command post, provide input at planning meetings, and advise the liaison officer of any special needs of his particular agency.

MEDICAL UNIT LEADER

The medical unit leader makes sure there are enough warm EMS personnel in the ambulance to care for the victims. EMS crews should also take the technicians' blood pressures before and after they work on the ice, monitor personnel for cold stress, and work the rehab station.

The responsibilities of the medical unit leader are to:

1. Obtain briefings from the IC or liaison officer.

2. Determine what first aid has already been administered, what is currently being done, and what still needs to be done, both immediately and afterward.

3. Determine whether sufficient medical personnel, supplies, and ambulances are on hand.

4. Prepare a medical emergency plan, complete with objectives   **49**
   and the procedures to accomplish them.

5. Ensure that the medical emergency plan is carried out.

6. Respond to requests for first aid for personnel, victims, family members, and bystanders.

7. Monitor personnel before, during, and after the operation.

8. Prepare documentation and a medical report.

Like the incident commander, the medical unit leader must look at the entire operation and be ready to address various problems that arise. Are EMS personnel who are outside wearing proper personal protective equipment? Where do EMS personnel meet the rescuers to take the victim? Should they be right at the shoreline, or should they wait at the outer edge of the warm zone? Is the transport device being used by the technicians capable of ferrying the victim to the ambulance? Is head or spine injury potentially involved? What hazards exist between the shoreline and the ambulance? How many personnel are needed to make the carry? Too many people trying to lift a patient over snow-covered rocks and roots could result in further injury if one or more of them should slip or trip over each other.

TECHNICAL SPECIALIST

A technical specialist is someone who may be outside the public safety industry but who can provide specific expertise relevant to the incident. Still, knowing who the right technical specialists are to call for advice or to perform a job is imperative if you are to prevent injury, further damage, and a waste of finances. An all-too-common example involves lifting a vehicle that has submerged under the ice. You'll hear about incidents in which a tow truck company was called to pull a vehicle to shore, the result being that the cable broke or the truck itself was pulled toward the water. Do ordinary tow truck companies have the training and equipment to do such jobs competently? In most cases, the answer is no. Instead, you should find a certified commercial diving company that can demonstrate the proper training, equipment, experience, and procedures to lift and stabilize the vehicle so that it can be transported to shore.

50   STUDY QUESTIONS

1. Who usually first assumes the role of Command?

2. Whatever the IC does not delegate is _____.

3. The command post should be set up in a location that is _____, _____, and _____.

4. True or false: The safety officer has the authority to cease operations if he perceives them to be imminently hazardous.

5. Keeping the public and the media in the cold zone, away from the IC and the operation, is the duty of the _____.

6. The link between the IC and other agencies is provided by whom?

7. Is it permissible for each responding agency to have its own staging area? Should each staging area have its own staging officer?

8. True or false: Besides overseeing the welfare of rescuers and victims, a medical unit leader is also obligated to respond to requests for first aid for bystanders.

6

Surface Ice Rescue Equipment

Before we can discuss how Awareness-, Operational-, and Technician-level rescuers can perform their duties, we need to look at some of the many types of equipment that they will be using. The key to choosing what to purchase is what your department can afford and the types of ice environments in which you may work. The average fire or police rescue team in the United States has virtually no budget for water rescue in general, much less its more exotic cousin, ice rescue. Some good news is that, when it comes to equipment such as transport devices, costlier does not necessarily mean better. Also, some tools such as ice poles and awls can be made at home at little cost. (Remember, though, to avoid making items that might cause injury if they fail. Legal liability is something that all departments should do their utmost to avoid.)

Is the ice rescue team at Operational or Technician status, and what level of competence does the county need? Are there other Technician-level teams already in existence? Perhaps the largest body of water in the county is only fifteen feet across and four feet deep. An Operational-level team may be enough to do the job, knowing that a mutual aid Technician-level team is close by. When making these decisions, keep in mind that unless a dive team is specifically trained in surface ice rescue, it isn't the same as a surface ice rescue Technician-level team. The procedures and equipment for surface and subsurface ice rescue are very different. If the local dive team is large, it may make sense to train them to the surface ice rescue Technician level. If it is a small team, this training isn't necessarily recommended, since a fully ready dive team should always be in standby status while a surface ice rescue Technician-level team works an operation. If the divers are working the surface operation, what happens when divers are suddenly needed in the water?

For a department in the process of developing an ice rescue team, it's best to train the members first before making significant equipment purchases. Good training will prevent teams from wasting precious funds on equipment that will be superfluous, that is inappropriate, or that isn't worth the cost.

Be creative when it comes to looking for ice rescue equipment if budget is a limitation. Check with Navy bases, for example, to see whether they are in the process of getting rid of PFDs. If the department absolutely cannot afford an ice board at this time, check out yard sales for an old surfboard that can be outfitted as an ice rescue transport device, or purchase a flotation backboard with at least 200 pounds of lift.

Possibly the most important point to remember, however, is not to become operational as an ice rescue team until all of the necessary equipment, training, drill time, and other requirements are completed and available. If you perform ice rescues with less, then your team will be expected to do so in the future, time and time again.

It is all too common for a fire department to dispatch untrained members to an ice rescue incident. These personnel try to make do by tying ropes around their waists and extending a metal ladder to the victim. The victim is saved and the only immediately obvious repercussions are a few hypothermic firefighters and an immersion-hypothermic, slightly banged-up victim. Suppose this same scenario is repeated a few more times over the years. Now some member of the department says, "Hey, we need ice rescue training and equipment, which will cost us about $5,000." The fire chief and treasurer only laugh and say, "Why in the world would we spend $5,000 for something that we've been doing just fine for no extra cost?" Sadly, such departments don't usually receive funds for training or equipment until one or more of their members are hurt or killed, or the family of a drowned victim sues the county for a failed rescue attempt.

Don't let your team fall into the same trap! Take the time to write out what you need to become an Operational- or Technician-level team, including estimated costs and proposed SOGs. Take the proposal to the decision makers and say, "This is what we need to perform ice rescues. If we do not have the minimum items high-lighted in this proposal, we won't be able to perform ice rescues as per our SOGs. We'll have to call in mutual aid, which will take an extra twenty minutes, significantly increasing the chances that any rescue will become a recovery."

It's been stated before: No commander should be able to tell members to perform an ice rescue without the minimum necessary equipment and resources.

OPERATIONAL-LEVEL RESCUE EQUIPMENT

Consider the following to be standard for the Operational level:

1. Class III or V PFDs of appropriate sizes for all shore-based personnel who will be in the warm zone. Class I PFDs are acceptable, but they're bulky.

2. Gloves, hats, face mufflers, coats, boots, antislip soles, and other clothing appropriate for the weather and general environment.

3. At least three rescue rope throw bags (75-foot length recommended).

4. At least one fire hose inflator and cap, with two 150-foot deployment bags, one nonlocking carabiner, and an SCBA bottle.

5. At least three harnesses, plus deployment tether line bags, straps, and carabiners if there are risky embankments in the district.

6. A reach pole with a shepherd's crook, length extenders, and a grapnel hook.

7. A messenger tag line or line-thrower gun.

8. A profile slate and writing utensil.

Personal Flotation Device

The most essential piece of gear for shore personnel is the personal flotation device. No one should be allowed within twenty-five feet of the waterline without one—no exceptions! If a slope leads to the water's edge, personnel must wear PFDs at the top of the slope. This rule may sound extreme, but consider what might happen if an on-site accident were to render someone unconscious. Without a PFD, that person could sink underwater in seconds, perhaps even unnoticed. Suddenly, the ongoing operation might be complicated by another search, rescue, or recovery. Even expert swimmers can drown in twenty seconds if the water is cold, if a head or spinal injury is involved, or if they are weighted down. Conversely, if a rescuer accidentally falls in the water while wearing a PFD, the entire operation doesn't necessarily have to stop. Also, although the rescuer will reflexively gasp on immersion, the PFD may keep his mouth and nose out of the water, preventing water aspiration, which could otherwise lead to panic and laryngospasm. The chance of his drowning will thus be significantly reduced.

A PFD reduces the chance of drowning for several other reasons as well, one of them being a decreased risk of immersion hypothermia. The more you work in the water, the more heat you'll lose. A person without a PFD will work hard to stay afloat. The PFD allows the person to hang or huddle motionless in the water, which is particularly important during the winter months. A properly fitting PFD also provides some insulation for the body's core. Since fatigue is another common cause of drowning, PFDs save lives by preventing the need for overexertion, thus decreasing the risk of leg cramps and subsequent drowning.

Two general rules apply to all PFDs. First, if the one you're wearing isn't properly zipped or buckled, you aren't wearing it. An open PFD will quickly come off in the water. Second, it must be the right size for you. It should comfortably keep your mouth well above the waterline and not significantly restrict breathing.

There are numerous styles of personal flotation devices, categorized by the United States Coast Guard into five different types. (The new European CE standard has four classes of PFDs, categorized as Buoyancy Aid, with 50 newtons of lift; Permanent Buoyancy Life Jacket, with 100 newtons of lift; Offshore Life Jacket, with 150 newtons of lift; and the Serious Life Jacket, with 275 newtons of lift.)

Type I PFDs: (Offshore Life Jackets, with a minimum 22 pounds of buoyancy for the adult size.) Type I PFDs are designed for offshore use, especially for rough conditions and where there is a possibility of delayed rescue. These bulky vests have the greatest flotation of all types of PFDs and are designed to turn most unconscious persons to a face-up position. They are made in two sizes: child (under 90 pounds) and adult. They are not preferred for ice rescue because their large forward bulk is a hindrance when extricating the wearer from an ice hole.

Type II PFDs: (Near-Shore Buoyant Vests, with a minimum of 15.5 pounds of buoyancy for the adult size.) These are similar to Type I PFDs, although they are designed for near-shore, calm conditions, where a rescue should occur quickly. They are less bulky than Type I PFDs and provide less flotation. A Type II vest will turn some persons to a face-up position. These PFDs are made in four sizes.

Type III PFDs: (Flotation Aids, with a minimum 15.5 pounds of buoyancy for the adult size.) These are designed for continuous, comfortable wear, thereby allowing the user to engage in activities such as water skiing, fishing, kayaking, and canoeing. Generally, they provide less flotation than a Type I or Type II PFD and won't turn the wearer to a face-up position. A person may actually have to tilt his head back to keep from going face-down. Type III PFDs may not have enough flotation to keep the wearer's face dry in rough waves. They are available in a variety of sizes and styles.

Type IV PFDs: (Throwable Devices, with a minimum of 16 to 18 pounds of buoyancy.) These are designed to be grasped and held by the user, and to be dropped or thrown to a person already in the water. They are best used in conjunction with wearable PFDs (Types I, II, and III, and some Type Vs). Type IV PFDs may be rings, cushions, or horseshoe floats. It's interesting to note that ring buoys were never designed or meant to be thrown to victims in the water. They were designed to be dropped into the water to victims who had fallen overboard from large vessels. The victim would then swim to the ring buoy while the ship made its turn to retrieve him. If a ring buoy is thrown and hits the victim in the head, it could be the final insult that drowns him. In addition, ring buoys don't deploy very far or accurately, especially if their lines are coiled. If you want to increase the effectiveness of ring buoy deployment, use a line bag attachment. Rope deploys faster, farther, and more accurately out of a bag than from a coil, as demonstrated by rescue rope throw bags.

Type V PFDs: (Specialty-Use Devices, with a minimum of 15.5 to 22 pounds of buoyancy for the adult size.) These are designed for specific purposes and may be used only for those purposes. Examples include whitewater vests, deck suits, and hybrid (inflatable) PFDs. Flotation coats and full suits are excellent for public safety personnel working near water in cold weather, since they offer complete waterproofing protection, insulation, and flotation.

Type III vest-style and suitable Type V PFDs are the preferred choice for public safety teams because they allow easy movement and provide sufficient flotation. Side-string size adjusters are an important feature for teams in which members must share PFDs. The size adjustment is also helpful to allow for changes in exposure apparel. Many of the PFDs available on the market have pockets and holders that permit them to be equipped with safety gear and tools. Some Type V PFDs also have built-in quick-release harnesses. In addition to equipping your team's life jackets with whistles, tools, and lights, you might also want to consider adding a logo or team name to the back of the jacket. This provides quick identification of the members at the scene. If the team has different-sized PFDs, clearly mark the sizes on the outside so that they can immediately be recognized, thereby saving time during the dressing phase.

Besides providing excellent flotation capability, flotation suits and coats provide waterproof protection from the elements, insulation from cold weather, rescuer visibility with international orange colors and reflective taping, and closures to decrease convective heat loss. Additionally, they afford the wearer excellent mobility for working. Some suits provide additional benefits, such as shoulder pockets for ice awls, hand-warmer pockets, added leg material for protection against wear and tear, and hoods with visors.

Purchasing PFDs: Should you purchase swift-water PFDs with built-in harnesses that release under a specific load? How about inflatable PFDs that are comfortable to wear all day long? Which is the best style: horseshoe, halterneck, vest, jacket, coat, or full suit? Are crotch straps necessary? How about built-in hoods, neck and face shields, rescue grip loops, or water-activated flashers? Should you purchase PFDs that can withstand the high temperatures involved with firefighting, since the members work a fireboat during the warmer seasons?

The first step toward answering the multitude of questions about PFDs is to write down all of the possible situations for which your department might need them. If your department performs swift-water rescue, then purchasing mainly swift-water PFDs makes sense. However, be careful with the harnesses, since they shouldn't be used for anything other than in-water swift-water rescue. In ice rescue, you don't want the harness to open under a high-tension load. When tethering shore personnel, use a standard water rescue harness.

The best types of PFD for surface ice rescue are the Type V, low-profile, vest-, coat-, and suit-styles. If the PFD is very bulky in front, rescuers will be forced to put too much weight on their hands to extricate themselves and the bulk of the PFD up and over the edge of an ice hole.

Turnout Gear and PFDs: Turnout gear shouldn't be worn near the water for incidents that don't involve firefighting, since without proper training, the wearer has a high chance of drowning if he becomes immersed. Every firefighter knows how heavy turnout gear becomes when it's saturated. If turnout gear is the only exposure gear that personnel have available, then they must wear PFDs of sufficient buoyancy. Test any proposed PFDs with turnout gear in a swimming pool to make sure they have sufficient flotation. Firefighters should wear their PFDs under their turnout coat, allowing them to ditch the coat if necessary. Also, few PFDs are capable of withstanding the high temperatures of firefighting. If you wear one outside the coat for fire-related incidents, it could melt.

PFDs and Surface Ice Rescue Suits: If a suit has its own inherent buoyancy, even when flooded, then no PFD is necessary. However, there are shell suits on the market that have no inherent buoyancy or insulation because they aren't made of neoprene. The wearer only has buoyancy as long as air stays in the suit and the dry-suit underwear is thick enough. If the suit leaks or floods, both the buoyancy and insulation are lost. For this reason, ice rescuers wearing shell suits must also wear PFDs unless they have on flotation dry-suit underwear.

Tools to Wear on a PFD

1. Water-activated flasher for low-light or night operations.

2. Cutting tools come in handy for a variety of reasons, from sharpening pencils to making on-site repairs. The most useful and effective cutting tools are paramedic shears. The wearers cannot hurt themselves with them as they could with a knife. Shears are also less expensive than most knives. Purchase good-quality shears that will hold up to water use and can cut through fishhooks and other debris.

3. A whistle or other sound-making device to alert personnel at a distance.

4. A timer or stopwatch to record the in and out times of the technicians and divers.

5. A small notepad and pen wrapped with eight to ten inches of duct tape. Sometimes the tender may need to make a note of something to tell the diver or team captain later, and the duct tape may come in handy for quick minor equipment repairs.

6. Tie wraps are useful, since it is sometimes quicker to tie a broken object back together than it is to get a replacement. Tie wraps come in handy for a variety of jobs.

7. A small flashlight or penlight, since sometimes a given task requires more direct illumination than is provided by flood-lights.

8. A centerpunch, which can be used to shatter car windows during vehicle extrication operations.

9. Ice awls.

Gloves

Gloves should stay on a rescuer's hands throughout an operation. It's common to see rescuers remove their gloves to tie knots or perform some other skill requiring dexterity. You should either wear gloves that will allow you to perform such tasks or wear heavy gloves or mittens with a liner glove underneath. Double-layer wool mittens are very warm, even when they're wet. Rubberized, lined gloves will keep your hands dry when you're handling wet line. Wool liner gloves work well because they keep hands relatively warm even when they're wet.

Emergency responses commonly involve rope evolutions and belaying techniques. Gloves that are suitable for handling lines must be available and worn during such operations.

Antislip Soles or Ice Cleats

Commercially made versions are available, or you can make them using tire treads and sheet-metal screws. In many cases, it's actually less expensive to purchase a good, long-lasting pair of anti-slip soles than it is to make a comparable pair. Either way, these are mandatory items.

Rescue Rope Throw Bags

The rescue rope throw bag is one of the most effective reach tools. It can reach as far as seventy-five feet, it deploys rapidly, and

Ice awls are useful for self-rescue, as well as for general mobility on the ice.

60 it floats. Also, it won't sink or damage the ice, and it won't likely hurt any victim who happens to get hit by one. It can easily be used from any type of embankment or vessel, and it is readily transportable to the place of deployment. Rescue rope throw bags are inexpensive enough for most departments to have several for multiple victims. Police and other first responders should always keep a rescue rope throw bag in their vehicles.

Commercially made varieties are available, or you can make your own. In either case, make sure the line is ⅜-inch floating braided polypropylene with a very flexible weave and some stretch. This won't absorb water and become stiff, and it has less of a tendency to freeze from the inside out as some other lines do. Braided polypropylene is relatively inexpensive compared with water rescue line. It's also durable and easy to maintain. Don't use heavy kernmantle water rescue line, because it won't deploy as fast, as far, or as accurately. Such line is static and a waste of money for this application. It's also less desirable because the victim, rather than the line, will take the brunt of being accidentally pulled into something. Test any bag before you make a purchase.

Never step on rope. Any line will weaken when you step on it, but this is especially true for wet, frozen line. After the operation, the line should be thoroughly cleaned, checked, and hung out to dry before you repack it.

To repack any rope bag, make an O with your index finger and thumb, and hold the opening of the bag with those fingers. Run the line through this O. Put your dominant hand inside the bag and pull the line into the bag. No coiling is necessary. This rapid packing method will prevent any knots, clumps, or line memory that could prevent full and rapid deployment of the line during the next throw. It can easily be done by one person standing or sitting, and it works well with gloved hands. With gloves on, a significant portion of any accumulated dirt can be removed as the line runs through the O.

Inflated Fire Hose System

To deploy a fire hose as an inflated line requires hose, a hose cap, an inflator, a carabiner, and two 150-foot deployment bags. Hose of 5 ½-inch diameter is easier to deploy straight, but the system would cost more. If you are unable to get a tag line across a given expanse of water for personnel on the other side to help you with deployment, the most efficient hose length to use is fifty feet. Anything longer will be difficult to deploy. As with throw bags, you

A fire hose deployed as an inflated line.

should use ⅜-inch braided polypropylene line for the deployment bags. Because an inflator hose, system, and cap can cost up to $400, some departments construct their own inflators and caps. If you choose to build your own, make sure the unit has a safe over-pressurization blow-off valve, and be aware of any liability issues in case failure and injury occur.

Water Rescue Harness

If the embankment is steep or particularly slippery, Operational-level technicians should be harnessed and tethered to immobile objects to prevent accidental immersion. The best method of tethering is to use a water rescue chest harness, which should be adjustable, so as to fit over different personal protective equipment. To don the harness, an assistant holds it open behind the tender, who either puts his arms straight back behind or straight up over his head. The assistant then slides the harness up both arms to the tender's shoulders. The assistant comes around to the front as the tender places his hand over his solar plexus region. The assistant closes the harness over the tender's hand, asking the tender to take a deep breath. This procedure ensures that the harness is not too tight, which could restrict breathing. This procedure also ensures that the harness sits correctly over the solar plexus region rather than the diaphragm or lower. If necessary, the shoulder straps should be adjusted for proper positioning.

A line or strap is then secured to the D ring on the back of the harness with a nonlocking carabiner. If the tender needs to move or be disconnected, a nonlocking carabiner will allow the quickest opening. A locking carabiner can freeze shut or be wrenched closed too tightly, stripping the screw, meaning that the tether line will have to be cut to save time. If the embankment is particularly risky, use two nonlocking carabiners facing opposite directions.

Reach Pole System

There are excellent commercially manufactured reach pole systems available. When used with various attachments, they allow rescuers to:

1. Pass off rescue flotation slings to victims.

2. Pass off rescue flotation slings to one victim during a multiple-victim incident while approaching another victim who is in a more critical state.

3. Perform animal rescue from a safe distance by using the shepherd's crook.

4. Pull a submerging victim closer to the technician by using the shepherd's crook.

5. Retrieve a submerged victim by using the shaft length extenders and a grapnel hook with rounded ends.

6. Use a boat hook adapter to fend off objects, hook into lines or edges, and push away drift ice.

Four sections of PVC piping make an effective, inexpensive reach tool. If it's made of metal, the pole should have a coated grip to protect the rescuer's hands.

You should get hands-on training to learn how to use the reach pole system for a variety of water-related incidents. When performing submerged victim retrieval procedures, add enough extenders to keep the rescuer's hands comfortably out of the water. When using the shepherd's crook in a current, position the bend of the crook downstream to catch the victim. To facilitate animal rescues, attach a lasso to the shepherd's crook.

Four sections of PVC piping make an effective, inexpensive reach tool.

Messenger Tag Line and Line-Thrower Gun

These tools can be used to set a line across a narrow body of water to deploy a rescue flotation sling, a ring buoy, an inflated fire hose, or a transport device directly to a victim. A good 150-foot messenger tag line can be purchased for under $60 and, with a little training and practice, can be used with excellent accuracy.

A line thrower, capable of sending line 300 to 1,000 feet, can also be very useful when you're confronted with a wider body of water. Use the line gun to set a tag line, which in turn can be used to pull across a heavier line, allowing shore personnel to move flotation aids and transport devices to the intended location.

Additional Operational-Level Equipment

Operational-level technicians may also want binoculars so as to keep a closer eye on the victim. Some kind of voice amplifier, such as a megaphone, will allow them to talk to both victims and rescuers.

TECHNICIAN-LEVEL RESCUE EQUIPMENT

A team can be fully dressed for Technician-level ice rescue with a budget of $3,000 to $4,000, which will provide:

1. Three PFDs.

2. Three ice rescue suits.

3. Three appropriately sized, adjustable water rescue harnesses with front and back attachment points.

4. Three aluminum locking carabiners and three nonlocking carabiners. Harnesses should have a locking carabiner in the front as well as a nonlocking carabiner in the back, so one pair is needed for each rescuer.

5. Two rescue flotation slings.

6. Two wooden ice rescue poles.

7. Three 250-foot marked line deployment bags or reels with braking systems for longer lines.

8. Two 75-foot rescue rope throw bags.

9. One ice rescue board or other transport device.

10. One quick-release contingency strap, which is placed between the strap on the ice board and the rescuer's front harness tether point. It allows the rescuer to disconnect quickly from the board if necessary. In any water rescue situation, you should never secure a rescuer to a victim without some means of releasing quickly.

If the budget allows, the following items would cost an additional $500 to $550:

1. Three pairs of ankle weights for ice rescue suits.

2. Three pairs of ice awls.

3. One pair of fins for an ice rescue suit.

4. Three shears and shear holders for harnesses.

5. Two ice anchor kits (piton, strap, steel carabiner).

6. Three pairs of antislip shoes.

7. Three ice rescue suit hood lights with hood attachments.

Water rescue helmets may also be needed, depending on the conditions and type of rescue.

Surface Ice Rescue Suit

Surface ice rescue suits are dry suits typically constructed of five- to seven-millimeter-thick orange or red neoprene, with attached

A surface ice rescue suit.

hoods, hard-soled boots, gloves, and a built-in harness. One of their drawbacks is that the tether points aren't in the best position for the wearer's body, and after repeated use, they can rip out of the suit. If you have these harnesses, remove them and use them as the straps for securing the ends of the tether lines to immobile objects. Some suits come with lobster-claw gloves that decrease rescuer dexterity, so five-fingered gloves are preferred, although they aren't as warm. Wool or fleece glove liners will help.

Do not purchase man-overboard survival suits, which are designed to keep people afloat and warm in rough seas. These suits have features that hinder ice rescue technicians. Face shields, for example, can be a liability. If your suits have face shields, you may want to remove them. Ice rescue typically involves a significant degree of exertion, so a technician cannot afford to have his airway covered with a neoprene flap that creates a stagnant, carbon dioxide-filled airspace. Also, ice rescue suits should have hard-soled boots with a tread that can be used with fins, antislip soles, or ice cleats. "Gumby" feet have more buoyancy than boots, so rescuers have more difficulty maneuvering with them, even while wearing ankle weights. Finally, ice rescue suits should move with the rescuer. If the suit is loose rather than form-fitting, then every movement will have to be exaggerated, and fatigue will set in more quickly. If you purchase shell suits, remember that they require winter-weight dry-suit underwear and PFDs.

Suit Maintenance: An ice suit's zipper is the most expensive part of the suit and can even be worth more than half the overall cost. Unfortunately, zippers cannot be fixed. Rather, they must be replaced, costing $150 to $225. The first step, therefore, is to prevent zipper problems.

Wax the zipper with paraffin at the end of every operation. Never use beeswax, which will hold dirt in and cause tooth damage when the zipper is opened or closed. If beeswax has been used, clean it out with a toothbrush. Do not use silicone- or petroleum-based lubricants, since these will degrade the neoprene over time. Buy a block of paraffin from a hardware store, chop it into smaller pieces, and place each piece in a clean, sealable plastic bag. Place one bag with each suit. Rub the paraffin on each closed zipper, and then open and close the zipper several times until it works smoothly and quietly. The only part that must be waxed in the open position is the very top that is under the zipper pull when it is fully closed. The paraffin will wash away during the operation and take the dirt with it, so reapply paraffin after every immersion.

The next areas of concern for suit maintenance are the suit seams. One of the main reasons surface ice rescue suits typically cost so much less than diving dry suits is their seam construction. Never put newly obtained suits into service without first checking for leaks. As many as twenty-five percent of some manufacturers' suits leak straight from the factory. It may be worthwhile to send one or two department members to an exposure-suit repair clinic for minor repairs.

To help prevent seam leaks, do not hang the suits, especially when they're wet and heavy, since doing so will stress the horizontal seams, which will slowly be pulled apart. After rinsing them, dry the suits by turning them inside out and, if possible, laying them on a hose rack or something similar. If that isn't possible, hang the suits at both the head and feet like a hammock. Once the inside is well dried, turn the suits right-side out, and again lay them out. Once they're completely dried, roll the suits with the zippers open and put them in a bag, with nothing placed on top of them. Keeping the zipper open decreases the stress on the zipper. Never

fold neoprene or put weight on it, since the creases will become
permanent weak areas.

If you're storing suits in a place harboring rodents, be aware that mice have been known to chew large holes in neoprene suits and leave grain and nut deposits in them.

Diving Dry Suits

If you have five- to seven-millimeter neoprene dry suits, use them for surface ice rescue. Those suits have inherent buoyancy and insulation, and they're generally less expensive than other types of dry suits. Vulcanized rubber, high-quality trilaminate, and crushed neoprene dry suits are too valuable and costly to risk damaging them during an ice rescue. Purchasing surface ice rescue suits will save money in the long run.

Water Rescue Harness

Technician-level rescuers must be in a horizontal position to work on the ice and to extricate themselves from a hole. Therefore, a water rescue harness should have a tether point at the height of the solar plexus, so that when tenders tension the line, the rescuer will be pulled into a horizontal position. If the tether point is lower, as on a rappelling harness, the wearer will be pulled into a vertical position, with the harness carabiner caught under the ice roof.

Ice Rescue Flotation Slings

A flotation sling is made of a loop of semirigid foam with one-inch nylon webbing that runs through its entire length. The foam provides at least 15 pounds of buoyancy, and the webbing provides the strength needed to extricate victims from holes. If the webbing doesn't go through the entire loop, it will probably tear out when under stress. A carabiner is attached to the end of the webbing to allow lines to be secured easily. Also, the sling has a self-fastening loop so that you can tighten the sling around a victim's chest.

Ice Poles

Wooden ice poles are among the most useful yet least expensive tools for the surface ice technician. They may be commercially made, as in a reach pole system, or they can be homemade. If homemade, make them of seven-foot-long hardwood banister poles: round, with one flat side, and of a size that may be gripped easily. Don't use pine, since it isn't as strong and tends to splinter easily.

Wooden ice poles are among the most useful tools for the surface ice technician.

Poles can be painted rescue orange or with alternating orange and white stripes. Also, at one-foot intervals along each pole, add wraps of good-quality duct tape to act as grips. Wrap enough tape so that each grip is ⅛ to ¼ inch thick.

Drive a three-inch nail to a depth of two inches into one end of the pole, then cut the head off. The nail then becomes a spike that can be used to grip the ice while walking.

In the other end of the pole, drill two holes about eight inches apart, and use them to insert a length of stiff polypropylene line, creating a loop. Tie knots on both sides of the holes to prevent the line from slipping back and forth.

Tether Line Deployment Bags or Reels

As with throw bags, use ⅜-inch braided polypropylene with flexible weaves. The line should float, and it should have some stretch. Line of a wider diameter, or heavier line, will present too much drag for both technicians and tenders. Static line will place unnecessary stress on the victim, technician, or transport device. Also, because drop strength isn't a real issue in surface ice rescue,

Tether line deployment reel.

there's no reason to waste expensive, high-load lines for these operations.

Lines should be kept in bags for the easiest mobility, deployment, and storage. If you use reels, they should have braking systems to control the speed of deployment to prevent the reel from outrunning the line, causing entanglements.

Line Markings: Line should be marked for distance so that you'll know when to add additional line, to mark the victim's distance out, and to know the technician's distance out.

Mark lines every five feet, with distinctive marks every twenty-five feet. The markings allow the chief tender to tell Command how far the victim is from shore, which should then be recorded on the profile map. Also, if the tether line is 250 feet long, for example, and the chief tender sees that the rescuer is 200 feet out and not yet at the victim, he'll know that he needs to call for a second deployment bag to extend the line.

Wrap a narrow piece of duct tape around the line to mark the five-foot increments. Refer to the accompanying illustration for

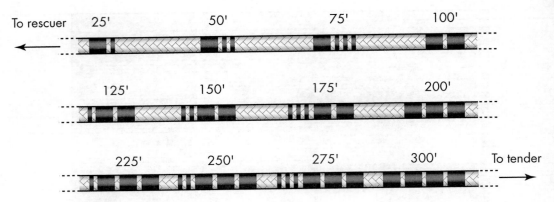

A tether line marking system.

line markings to designate actual distances every twenty-five feet. If possible, use different-colored tape for each hundred feet. For example, use green for the first hundred because it is the safest, and use red for the 300- to 400-foot marks because that designates a long-distance operation requiring additional training, equipment, and personnel. When the rescuer is on the tender's side of 100 feet, the narrow pieces are on the tender's side of the 100-foot mark. When the rescuer is beyond 100 feet, the narrow pieces are on the rescuer's side of the 100-foot mark. At 100 feet, there's double the effort; at 200 feet, there's triple the effort, and so on.

Use nonlocking carabiners to attach a tether line to webbing, which is then secured to an attachment point. If the tether line deployment bag location must be moved or extended by attaching a second line bag, rescuers won't have to contend with a frozen, locked carabiner. Long-distance operations will require line reels with 300 to 900 feet of line, pitons, and ice rescue suits for tenders and additional rescuers. Make sure that the line reels have brakes so as to prevent entanglements that would bind up the reel.

Long-distance operations of 300 feet or more from shore require support personnel on the ice. These personnel should be certified to the Technician level, be tethered, and be wearing proper personal protective equipment.

Transport Devices

One of the keys to short rescue times is having the right transport device. The right transport device decreases the amount of time needed for access, extrication, and transport to shore. It also

Ice rescue boards.

decreases rescuer exertion and the chance of victim injury. Given the variety of sleds, kayaks, ramps, and other ice transport devices available, as well as conditions ranging from open water to broken ice, and weather ranging from calm conditions to full-out blizzards, the ice rescue board is one of the most effective ice tools available. The inflatable Class V whitewater kayak is also excellent. The advantage of the kayak is that it's usable for rescue all year long. However, it won't travel across the ice nearly as well as an ice board. Whatever other considerations come into play when making purchasing decisions, if one rescuer cannot pick up

72   a transport device and deploy it by himself, you should think twice about buying it. Bigger isn't necessarily better. In practical terms, bigger usually means heavier, meaning it requires more personnel and time to deploy it, and it's more likely to break weak ice. Surface ice rescue requires light, quick rescue capabilities. Surface ice rescue needs to be completed on the first attempt. Also, rescue tools that have a high profile and are prone to being blown by wind aren't usually a good choice, especially if you need to cross areas of open water. If it cannot easily be transported over open water or slush, the item won't be very useful.

Reconsider your purchase, too, if the rescuer cannot easily establish buoyancy for the victim while using the transport device for support. If technicians will strain their backs placing the victim on the device, also reconsider. The device should not place pressure on the victim's diaphragm and result in a loss of airway when the victim becomes unconscious. Ask whether the transport device is suitable for victims with spine, head, and other injuries. If the device cannot be passed off to an aggressive or self-rescue victim to allow the victim to climb onto it, there may be a better purchasing option.

As mentioned above, inflatable Class V kayaks are effective ice rescue tools and are very stable for year-long surface operations and as dive operation support tools. They are in the same price range as ice boards. Hard kayaks are more difficult to use and can be less gentle for the victim than inflatable kayaks, but they're less than half the price. Properly rigged flotation backboards can also be used. They don't work nearly as well as boards, kayaks, and some sleds, but they can be purchased and properly rigged for under $200.

Inflatable platforms and walkways are excellent tools when used properly for certain conditions. They're incredible platforms for muddy or swampy areas, but may not always be the best option for ice rescue. Two ramps are needed to move across the ice toward a victim. The procedure is to pick up one ramp and place it in front of the other, leapfrog-style. Moving ramps across the ice this way requires at least two members, and the process can be time-consuming and exhausting, since the ramps are relatively heavy and cumbersome. Also, a single ramp is generally more expensive than other transport devices. Most aren't self-bailing, so the rescuers must work to keep water out and away from the victim once he's inside of it. Ramps must be inflated before use, which may delay a rescue, and they're more susceptible to wind

Inflatable kayak.

than ice boards. Still, a platform provides for a more stable transport than most boards and sleds. Platforms can be of great aid when spanning water to reach ice, and they're useful when crossing some moving water. All in all, if most of your ice rescues occur less than fifty feet from shore, then a fifty-foot ramp is an excellent tool to have in your inventory.

There is no one perfect tool for any job. This age-old axiom applies to ice rescue as much as to any other endeavor. Make sure that you try out a piece of equipment under a variety of conditions before purchasing it. Try out several different types before making a final decision. Once it's in your inventory, learn how to use the equipment you have at hand and become proficient with it.

With several new transport devices on the market, it's important that teams understand the necessary features and functions so that they can choose the device that's best for them. This is especially true since a transport device is the most expensive piece of ice rescue equipment that a team will purchase.

A transport device should be:

1. Safe and effective in a variety of conditions, including open water, breaking ice, slush, good ice, and snow-covered ice.

74

2. Minimally susceptible to the wind; i.e., it should have a very low profile.

3. Stable and not easily flipped by an aggressive victim.

4. Easy to right if it is flipped over.

5. Lightweight.

6. Easily portable, so that one rescuer can easily carry it through brush and over terrain to the rescue site.

7. Simple to use, since most departments conduct fewer than four ice rescue drills annually.

8. Made of nonmetal materials to prevent the great conductive heat loss that occurs from contact with cold metal. Metal can also become very slippery when it's wet, increasing the potential for falls and handling difficulties. Wet skin can also freeze to metal if the temperatures are cold enough.

9. Designed with rounded corners and foam-protected edges to prevent pain and injury should the victim inadvertently bang into it.

10. Safe and effective for all types of victims, whether aggressive or not. The device should allow an aggressive victim to mount it alone, without direct help from the rescuer, thereby protecting the rescuer. It should also provide immediate buoyancy to the victim.

11. Compatible with safe, gentle handling of the victim. It should allow the patient's airway to be maintained in case loss of consciousness occurs. Horizontal extrication must be possible with it so as to keep blood in the core of the body. It should allow for stabilization of the head and spine, as well as transportation with splinted extremities. It should also be able to serve as the backboard transport device from the shore to the ambulance, thereby decreasing the number of times that the patient must be transferred to a new device. It should also be a viable platform for CPR and other emergency procedures.

12. Storable in vehicle compartments.

13. Affordable.

14. Durable.

Hard kayak.

Flotation backboard.

Short inflatable ramp.

15. Useful for conveying equipment and personnel for dive operations.

16. Capable of serving as a tender platform on weak ice.

Long-distance operations require self-powered transport tools that can cross possibly miles of ice and open water. One option is an air boat. Operators of these craft require extensive training and continued practice. Any department using such a vehicle should also be prepared for relatively frequent repairs, which can be expensive. Rescuers also need to learn how to load a patient into the vessel, as well as how to protect both victims and rescuers from the windchill effect of the fans. Some teams have discovered that it's necessary to put protective material over the air bags, which have been known to puncture when the vessel is loaded with a heavy patient and it travels across chunks of broken ice. Another consideration is that some smaller models force one rescuer to remain behind on the ice for later pickup because there's no room for the patient and a second rescuer.

Roll-Up Straps

It is extremely difficult to get any victim into a transport device or boat, much less someone weighing more than 200 pounds. Bringing a victim up into a vessel takes more than just practice—it takes technique and equipment. The roll-up strap technique employs both reduced body weight and leverage. To make a set of roll-up straps, take a pair of two- to three-inch flat webbing straps approximately thirteen to fifteen feet long, and put a No. 4 grommet in one end of each one. Then attach standard carabiners to each grommet.

You can now place and secure the open end of the carabiner under the lip of a vessel's gunwale. Actually, the carabiners can be secured to anything on the inside of the boat, and, if necessary, personnel can even stand on the carabiners or the ends of the straps. The length of each strap will now go between the victim and the boat, then be passed under the victim, around him on the outside, and back up to the rescuer's hands.

Place one of the straps midway between the victim's elbow and shoulder, and place the other strap midway between the victim's knees and hips. Put the victim's arms inside the straps. The straps can be set around the victim by personnel in the vessel or by a rescuer in the water.

Although one rescuer can pull up a full-sized adult, it's much easier with two rescuers. It is imperative that the straps be checked before they are pulled. As a strap is returned to the boat after coming around the victim, it must lay directly over itself, over the first part of the strap first leaving the boat. If the strap returns to the boat at an angle, it will attempt to straighten itself out as the victim is rolled into the vessel, with the result that the strap will move on the victim. If this occurs, the strap could end up around the victim's neck. Next, check that the straps are fully perpendicular with the gunwale. Again, if they're at an angle, they will straighten out and the victim will either be hanged or he'll fall out. If a rescuer is in the water, then he is the best person to check the placement of the straps.

After ensuring that the straps are properly in place, the person in charge counts to three, and both rescuers simply roll the victim into the vessel by pulling in on the straps. The rescuers should stay low in the boat, continuously counterbalancing the vessel to prevent it from tipping or flipping. The person nearest the victim's head must maintain control of the procedure. If the lower body is raised too fast, it becomes difficult to get the victim into the boat.

If the upper torso is raised too fast, the shoulder strap can work its way up to and around the victim's head and neck. Once the victim is about to reach the gunwale, the head rescuer reaches over and secures the victim's head so that it doesn't hit the boat and so that an airway is maintained. The victim is then laid down gently into the vessel, onto the deck, or onto a transport device.

Lay a backboard on the gunwale if the gunwale is too narrow to safely support the victim. Then lay the victim directly on the backboard and secure him. Use a flotation backboard with enough lift to float the victim in case the boat flips. Store the straps rolled up and secured to the gunwale with one wrap of duct tape. Fold a tab onto the end of the tape for rapid release.

Other tools can be used to achieve the victim roll-up strap procedure, such as cargo netting, a ladder of webbing, and wind fencing. Whatever you use, first train with a boat on land, then practice in the water. Make sure that rescue personnel are wearing the same gear and gloves they would be wearing during an ice rescue.

Roll-up straps can also be used to right an overturned boat in open water. If a vessel overturns, rescuers in the water can simply reach under the gunwale, release the straps with one end still secured, and throw the straps over the hull. They then climb up onto the opposite side of the hull, then stand with their arms and backs straight, holding the straps. Leaning back, their weight will turn the boat right-side up.

LAST-RESORT RESCUE TOOLS

Metal Ladders, Poles, and Boats

In water operations, metal objects become cold very quickly and will conduct heat away from any warm object, including your hands and body. They may also sink unless you have attached some sort of buoyancy. Like other items made of metal, they tend to freeze to the ice. Consequently, they aren't the most preferred items for any water-related rescue, particularly those on ice. Besides these drawbacks, metal boats aren't very stable in open water and become less so on the ice. They become quite slippery during rain and snow and are difficult to transport across the ice. Usually they require that the ice be broken up. With a metal boat, it's also very difficult to extricate a victim safely from an ice hole and move him horizontally into the vessel.

In short, metal rescue tools aren't usually the best options for ice rescue.

Winches

Resist the urge to use a winch whenever a human being is on the other end. A mishap can result in severe injury or death to the person being pulled or anyone standing near the line. Also, winches can create a static line, especially in the case of wet, frozen line, which is more likely to break. Such a precaution may seem like common sense, but given some training programs, drills, and actual rescues, apparently it isn't.

80   STUDY QUESTIONS

1. Without exception, no shore personnel should be allowed within twenty-five feet of the waterline without a _____.

2. The two types of PFDs that are preferred for public safety teams are Types _____ and _____.

3. True or false: If a surface ice rescue suit has its own inherent buoyancy, even when flooded, then no PFD is necessary.

4. Standard Operational-level ice rescue equipment should include how many rescue rope throw bags and of what length?

5. What items are required to deploy a fire hose as an inflated line?

6. True or false: Generally, a locking carabiner is preferred over the nonlocking type when tethering a harness for ice rescue operations.

7. A line thrower gun is capable of sending line approximately how far?

8. To maintain its performance, you should wax the zipper of an ice rescue suit with _____ at the end of every operation.

9. The tether point on a water rescue harness should be at what point on the body so that the wearer will be pulled into a horizontal position?

10. Tether lines should be marked every _____ feet, with distinctive markings every _____ feet.

11. Name some of the features of a good transport device.

12. Besides their usefulness in bringing a victim aboard, roll-up straps can also be used to _____.

7

The Rescue:
Awareness-Level
Duties and Responses

On arrival at the scene of an incident, Awareness-level personnel should:

1. Determine the best place to stage, based on predetermined sites and the victim's believed location. Determine whether special transport devices, such as snowmobiles, are needed to reach the location. If so, make sure that they are dispatched.

2. Don a personal flotation device and other necessary personal protective equipment, including a hat, gloves, and boots. If the shore is icy, wear ice cleats as well.

3. Assess the scene and identify hazards, such as where the ice begins, slippery embankments, obstacles in the ground, downed power lines, leaking fuel, and the like. Notify the appropriate agencies to deal with these hazards, if necessary.

4. Obtain preliminary information from witnesses. For example, how many victims are there, and how did they get there? How long have they been there, and what is their status? Also, ascertain whether any witnesses or bystanders are in need of medical attention.

5. Mark the spots onshore in front of the victims. As a first responder, you should have an available means of doing this rather than waste time trying to find markers.

6. Call for the appropriate agencies. Always dispatch a dive team to support surface efforts as well as cases of submersion. If at least one victim has submerged, notify the local hyperbaric facility that divers will be entering the water. Not all facilities are staffed twenty-four hours a day, but hyperbaric technicians and physicians may be on standby duty. If the closest hyperbaric chamber is far away, air transportation may be required, so plan a landing zone. Ideally, you should have an ambulance for each victim, plus one or more for rescuers and bystanders. Part of maintaining scene safety is to make sure no one becomes hypothermic or injured in other ways. You may need to notify the ambulance personnel that proper exposure wear is necessary. When on call, EMS personnel shouldn't forget to take their hats, gloves, appropriate clothing, and boots—seemingly a matter of common sense, but countless incidents prove that it isn't.

7. Establish command and initiate the IMS. Designate a command post, staging areas, staging officers, a safety officer, and other sections as required by the size and complexity of the incident.

8. Secure the area. Make sure that no unauthorized persons go out on the ice. Establish hot, warm, and cold zones. Secure all of the witnesses and, if possible, turn them over to law enforcement personnel for interviewing. If no police are available, then use Operational-, Technician-, or Awareness-level ice rescue personnel, if available. Also, determine where the wind is coming from, and direct incoming vehicles to turn sideways to it to form windbreaks for the staging area to protect personnel as much as possible. Stage the arriving ambulances for easy access and immediate egress. Plan and direct a protected staging area for the dive team as well. In directing vehicles, be sure to maintain access and exit routes for fire and EMS vehicles.

9. Draw a profile map of the victim's location for the responding surface-rescue and dive teams in case the victim submerges or nightfall arrives. Find a safe place to deploy ice rescue technicians, as well as a shore point for the most direct route from the hole to the shore.

10. Reassess the status of the victim, if possible. If the victim is becoming passive or is close to submerging, notify the dive

team for rapid deployment. Know what the maximum submersion time is for the operation to be in rescue status, as denoted by the SOGs. Reassessment is also very important for multiple-victim incidents so as to determine who needs to be rescued first and what types of effort are required.

11. If Operational- or Technician-level ice rescuers haven't yet arrived on the scene, begin communicating with the victim, if possible.

12. If higher-trained ice rescue personnel haven't yet arrived, interview the witnesses in more detail and document the information on the profile map.

13. Do not go out on the ice.

INTERVIEWING WITNESSES

Interviewing witnesses and drawing profile maps can be considered Awareness-level duties, since they're shore-based activities. Training in interviewing and profiling at the Awareness level is optional, but it's mandatory at higher levels. Because both require greater training, they should be turned over to someone trained to Operational or Technician levels whenever possible. Fortunately, law enforcement personnel, who should already have interviewing and recording skills, are usually present.

The first step in interviewing witnesses is to acquire two critical tools: paper and a writing utensil. Without writing the information down, you'll never remember exactly what the witnesses said, nor will you be able to relay it properly to someone else. Moreover, you won't know in a second interview whether you've received the same information or not. Without a written record, you won't be able to correlate information from different witnesses, and you won't be able to determine the victim's location to as high a degree of certainty.

Information gained from witnesses is especially important if the victim cannot be seen or heard, as in the case of long-distance operations, low-light or storm conditions, or after victim submergence. An interview may also provide information on the length of immersion, possible trauma or other medical problems, the number of victims, and the presence of additional hazards.

Witnessing an accidental immersion or drowning can be emotionally traumatizing. The mind sometimes shuts down from such painful experiences by suppressing or blocking out memories of the event. This may be especially true if feelings of guilt and helplessness are involved. Guilt can also cause the witness to imagine or create information to "be more helpful." For these reasons, the body language of witnesses may provide more accurate information than the literal answers they give. The body has a memory as well, and once tapped, it can sometimes reawaken the blocked mind's eye. Body language itself can often provide valuable information on the veracity of a witness's statements and of the victim's location.

To gain information from both body memories and the mind's eye, ask each witness separately to reenact exactly what he did immediately prior to and during the incident. If a witness states that he ran to the end of the dock, ask him to do so. Document the information, then go over it with Command to see whether it's logical. Pay close attention to where the witness directs his body when he reenacts what he did during the incident. Where he faces and places his body may provide more accurate information than where he points.

Some examples of questions to ask witnesses:

- What alerted you to the incident?
- Where were you when you first saw or heard the incident?
- What where you doing?
- Can you please reenact exactly what you did?
- Were you alone?
- Was the victim alone?
- Where was the second or third victim (in the same hole, forward, or behind)?
- How long ago did this happen?
- What color clothes was the victim wearing? This could be very important if, for example, a red coat is floating and the victim was reported as wearing a blue coat.
- In what direction was the victim moving? Where was he coming from or going to? The victim may have walked across the best ice, which may be the better route for the rescuers to take.

Children often make excellent witnesses, because they typically don't lie or create information as adults might. Unfortunately, if adult witnesses are available, children are often overlooked. **85** Animals have also been known to indicate where their owners are, whether on the ice or under the water. Don't limit your witness options. You may need to contact friends or family who weren't on the scene to provide more information about the missing person's habits. Some typical questions might be: "Where did your friend normally fish from on the lake?" "Have you ever seen your child playing on the ice?" "With whom does your child normally play after school?" "What color coat did he wear today?"

PROFILE MAPS

Profile maps are critical for planning and searching for a missing person who may have ended up on the ice. As stated earlier, the first-arriving responder should quickly create a profile drawing of the scene, showing exactly where the victim is, based on either visual contact or witness accounts. This information is very important for showing other rescuers where the victim is, and it becomes critical if nightfall comes; if the victim submerges and the hole freezes over; if the victim submerges and rescuers break up the ice, thereby destroying the hole; or if the weather decreases visibility.

Profiles and other documented information can also be valuable for debriefings, postincident analysis, training, drills, and legal disputes.

A drowning recovery involving ice can be relatively simple for dive teams if they are properly trained and receive accurate information as to the victim's location. If a dive team is responding, surface teams should avoid destroying the ice so as to avoid destroying the hole. It takes skill and training to draw a very accurate profile map, and just as much training and skill to deploy divers using that map. If the hole has been destroyed, it becomes that much more difficult to deploy the divers with any accuracy. If the dive team is en route, it would be better either to wait for them or to have a tethered ice technician lie on the ice by the hole to search the bottom with a pole. Avoid using reach poles with sharp hooks that could injure the victim. If possible, drop a marker buoy in the hole if the victim submerges, but avoid destroying the ice.

The ice puncture gives the ice technician or dive team an advantage that they don't have in any other drowning situation: a last-known site marked on the water. Drowning victims in nonmoving water are usually within a lateral distance from the hole equal to the depth of the water. Given a depth of twenty feet, the victim will most likely be within twenty feet of the hole. This will vary if the victim submerged completely on falling through the ice and then raced around under the cap trying to find a way out.

In moving water, consider both the speed and depth of the water. A body drops at approximately two feet per second—a little faster in freshwater and a little slower in salt water. Calculate the water speed and multiply that by the time it would take for the body to reach the bottom. This will give you the approximate lateral distance of the victim from the point of descent.

Contrary to popular belief, once a body reaches the bottom, it will stay there until gases form within it and gently lift it up again. Some teams acquire the belief that bodies move across the bottom, since bodies are sometimes recovered on the surface miles from the descent point. The key is that bodies don't immediately pop up to the surface once they lift off the bottom. They may travel underwater for miles before reaching the surface. Until they lift, however, they'll remain where they are unless the water is very shallow and moving extremely fast. In very cold and deep water, the body may never lift at all and will thus remain in the same location.

Drawing a Profile Map

1. Use full-size paper (8 ½" X 11") or a slate of comparable size.

2. Draw the shoreline, including landmarks, the point where you're standing, the location in which the witness was standing, and the marked shore location directly in front of the victim.

3. If visible, draw the opposite shore with landmarks and ranges (see below).

4. Draw where the victim is or is believed to be in relation to the ranges.

5. Estimate and write down the distance from the victim to the shore.

6. Include your name; the names and contact information of any witnesses; the number of victims; the name, description, and estimated time of immersion of the victim; the current time; and anything else required by your department's SOGs.

If possible, it's also useful to record the time that the rescuers set foot on the ice, the time that they reach the victim, and the

time that they bring the victim back to shore. Make note of the **87**
number of personnel involved, what equipment they use, and the
procedures they implement. Draw the route that the rescuers
take to the victim, as well as the egress route of transport. If any
rescuers fall through the ice, document that as well.

Ranges

To find a range, look at the target object, such as the victim or
the hole in the ice. Then find two objects that line up directly
together, preferably immediately behind the target. Look to your
left or right and find two more objects that line up perfectly, one
behind the other. Walk away, then come back to the spot and position yourself so that all of the objects are realigned as they were
before. If so, then you should be standing in the exact spot you
were in before.

ASSESSING THE VICTIM

The person who assesses the victim is called a spotter. The spotter keeps his eyes on the victim or the victim's last-seen location
at all times. The spotter reports any change in the victim's location or status. The spotter is often the communicator as well, and
he may need to be equipped with binoculars and a mechanical
voice aid, such as a megaphone.

The acronym ASAPS offers five criteria for assessing the victim:

A: Is the victim *aggressive?* Does he cooperate with verbal
   commands?

S: Is the victim capable of *self-rescue?* Can he be talked out of
   the hole?

A: Is the victim *alert?* Is he attentive to verbal communication?

P: Is the victim *passive?* Does he give any type of response to
   verbal commands or questions?

S: Is the victim *submerging?* What are his chances of remaining at the surface? Does submergence seem imminent?

Aggressive: An aggressive victim is one who is trying to pull or
fight his way out of the hole with little or no success. He is using
too much energy and breaking the ice that's supporting him.

If the victim is responsive to your verbal communication, find
out whether you can slow him down and perhaps even talk him

88 out of the hole. Try to change his behavior so that he becomes capable of self-rescue. The objective is to prevent further energy loss and destruction of the ice. His exertions will only result in more heat loss, fatigue, and a higher chance of drowning.

Rescue technicians must be prepared to manage and possibly defend themselves against uncooperative, aggressive victims. Since alcohol is often involved in water-related incidents, rescuers may be faced with intoxicated victims whose behavior could frustrate the rescuers, causing their tempers to flare. Rescuers must always stay calm and do their best to communicate effectively with even the most difficult of victims.

Self-Rescue: These victims may be capable of helping rescuers or even of rescuing themselves. They may initiate dialogue and be asking you to tell them what to do. If a victim gets himself out of the hole, ask him to roll slowly away from it and then to perform several more slow rolls. If the ice is weak, have him wait there for approaching ice rescue technicians. If he were to fall through the ice again while attempting to make it all the way back to shore, he might very well not be able to get himself out a second time.

If a victim can't extricate himself after several attempts, have him save his strength for holding on to the ice and working with the approaching rescuers. Reaching-and-throwing Operational-level rescue efforts may prove very successful with self-rescue victims who are close enough to shore for such procedures.

Alert: These victims are communicating and listening but cannot perform a self-rescue, either because the ice is so poor that it simply won't support their weight or because the victims no longer have the strength to help themselves.

Victims in any of the above three categories may be capable of:

- Verbal communication.
- Grabbing a rescue rope throw bag.
- Receiving and using ice awls, jumper cables, or other devices to assist in self-rescue.
- Receiving and using a rescue pole.
- Assisting in a tethered rescue flotation sling or ring buoy rescue.
- Assisting in an inflated fire hose rescue.
- Assisting in a sled or platform rescue.

Rescue technicians must be prepared to manage and possibly defend themselves against uncooperative, aggressive victims.

Passive: These victims aren't struggling, nor are they responding well to your questions and commands. They don't appear to hear you. They don't seem very active, attentive, or alert. Passive victims may make repeated vocalizations, such as a child crying for his mommy.

With effective communication, passive victims may be capable of:

- Verbal communication, which may not be coherent.

- Grabbing a rescue/throw bag, although a victim may not be capable of holding on to it.

- Receiving and using ice awls to assist in self-rescue, although a victim may not have the strength in his hands to use or hold on to them.

- Receiving and using a rescue pole, although a victim may not have the strength to hold on or to pull himself out.

- Assisting in a tethered flotation sling rescue or a tethered ring buoy rescue, although a victim may not have the strength to hold on.

- Holding on to an inflated fire hose.

- Assisting in a sled or platform rescue.

Submerging: These victims are drowning. Drowning is a silent event. Their heads may seem as if they're bobbing up and down slowly at the surface. When their mouths reach air, they are only capable of gasping for breath and cannot vocalize. Adults take an average of sixty seconds to drown, while children take an average of twenty seconds.

Submerging victims look as if they're just barely hanging on to life. There is no response to your verbal communication, and they look as if they could go fully under any second. Wide-eyed, they will neither see nor hear you. They may have instinctive reactions to the rescue attempts but cannot be trusted to hang on or even to comprehend what is taking place. Submerging victims won't be capable of any of the above actions or responses, so you can't expect them to give you any assistance.

Unless you immediately establish buoyancy for the victim, the rescue will quickly become a subsurface operation.

ESTABLISHING COMMUNICATION

Communication with victims is imperative for their survival. If the victim's status allows it, two-way communication has the greatest benefit toward survival. As mentioned above, the spotter is often the person assigned the duty of communicating with the victim. He should make visual and verbal contact with the victim if possible, using a microphone or bullhorn if necessary, but he should not go out onto the ice without Technician-level certification or equipment. Use the ASAPS information to determine the most appropriate means of communication.

If possible, find out the victim's name, and use it often. A two-way dialogue is the most effective. If you can keep the victim talking to you, he'll have a better chance of staying alert and oriented. Ask him basic, personal questions, such as his name and age, as well as anything else that will keep him focused.

Ask the victim whether he was alone, or whether there is another potential victim. Ask whether anyone tried to help him earlier. Perhaps a freelance rescuer or bystander made an earlier attempt at rescue and is now underwater.

Binoculars may aid communication if the victim is far from shore. If you cannot hear the victim's verbal responses, ask him to shake his head slowly for a negative response and to nod slowly for an affirmative response. Avoid having him use arm-raising movements, which could end up breaking supportive ice as all of his weight is shifted to the other arm.

Your communicating with victims may mean the difference between life and death. Whenever possible, do not use more than one communicator at a time, and retain the same person throughout the operation. Stranded victims are likely to develop trust with the first authoritative person with whom they communicate.

At the same time, prevent the victims from hearing their friends and family panicking. If the victim is a child, and if the parent can remain calm, then you may enlist the parent's help in communicating.

If the victim is aggressive, assure him that help is on the way. Ask him to slow down. If he is overly aggressive, command him to stop any struggles or futile efforts to climb out of the hole. Instead, have him rest his arms on whatever ice is left and to keep his face out of the water, wasting as little energy as possible. If he can speak to you, try to establish a two-way personal dialogue with him. As mentioned above, try to determine whether anyone else was involved in the incident. By maintaining back-and-forth communication, such a victim may become capable of self-rescue.

Once it has been determined that a victim is capable of self-rescue, give clear, simple commands to help him get himself out of the hole. His moves should be made slowly. Keeping his face near the ice, he should stretch his arms forward and kick gently, wriggling out of the hole and onto the ice like a snake. At no point should he rise up on his elbows. Once out of the hole, he should roll away slowly, distributing his weight across the surface, never rising up. If possible, have the victim roll toward approaching rescuers. If that isn't possible, have him lie still in a place of relative safety until rescuers can establish buoyancy for him.

If the above techniques don't work, toss out ice awls or other self-rescue tools, and instruct him on how to use them. Try other operational rescue efforts as ice technicians prepare and deploy.

Of course, if the attempted rescue is shore-based, you must instruct the victim on how to help himself. However, if any such attempts appear to be failing, if the ice continues to break, or if the victim is quickly becoming fatigued, you must terminate the self-rescue efforts and tell him just to hold on and stop moving.

If the victim is attentive and alert, maintain reassuring two-way communication. Since self-rescue efforts have been ruled out by circumstances, the best thing to do is to keep him alert and still. Have him keep his face out of the water and wait for help to arrive.

92   As long as he responds to your prompts, the lower his chances will be of slipping into the passive or the submerging state.

Even if a victim has become passive, you should continue talking to him. Communication is imperative, because hopefully he is still hearing you even if he isn't responding. Communicate loudly and forcefully to keep him as alert and oriented as possible, and try to elicit verbal or physical responses.

If the victim is submerging, trying to communicate will do little, since he probably won't even be able to hear you. However, you should still not abandon your efforts to communicate with him. Loudly and repeatedly command him to lie on his back and float. The sound of your voice may be enough to make a difference in his chances for survival.

STUDY QUESTIONS

1. True or false: Although they are shore-based activities, the Awareness-level tasks of interviewing witnesses and drawing profile maps should be turned over to Operational- or Technician-level personnel whenever possible.

2. If a witness is particularly traumatized, what might provide more accurate information than the verbal answers he gives?

3. To gain information from both body memories and the mind's eye, what should you ask each witness to do?

4. Why do children often make excellent witnesses?

5. The primary purpose of a profile map is to _____.

6. A drowning victim in nonmoving water twenty feet deep will probably be within how many feet of the hole?

7. A body drops at a rate of approximately _____ in water.

8. The person who assesses a victim is called a _____.

9. What are the five categories of victim assessment as indicated by the acronym ASAPS?

10. By establishing a two-way personal dialogue, an aggressive victim may become capable of _____.

8

The Rescue: Operational-Level Duties and Responses

When responding to an incident, an Operational-level technician should:

1. Communicate with on-site agencies while en route. Find out how many victims there are; what their ASAPS status is and whether it is changing; the basic condition of the ice; how far from shore the incident is; whether vehicles are on the ice; and any other pertinent information. If possible, dress while en route.

2. Determine on arrival whether the incident management system is in place. If not, institute it. Obtain whatever information has been recorded by the first responders, and ensure that adequate resources are on-scene, en route, or on call.

3. If there are not enough rescuers present to save each victim at once, you must triage the victims. Those in a passive and submerging condition, as well as those experiencing trauma or other medical emergencies, are at greatest risk.

4. Plan, initiate, and maintain appropriate victim-rescuer communication. In the case of a submerging victim, the plan could include throwing a marker buoy into the hole for possible subsurface rescue efforts. If no dive team is available, initiate reach-pole procedures to search for a submerged victim.

OPERATIONAL-LEVEL RESCUE PROCEDURES

Reach Method

The reach method entails deploying a tool that establishes a direct link between the victim and his rescuers. It is used when the victim is close enough to shore and the rescuers can be properly secured to an immobile object there. The purpose of the link can be to establish buoyancy, pull the victim toward shore, or provide a path for the victim to follow. If an object is used as a pathway to shore, it should be buoyant so that it will support the victim if the ice breaks. It should also spread the victim's weight over a large surface to decrease the stress on any one section of ice. Examples of such links include inflated fire hoses, floating backboards, life ramps, ice boards, inflated ramps, and buoyant wooden ladders.

If possible, avoid reaching out with nonbuoyant or metal objects unless there are no other options. A nonbuoyant object could weaken or destroy the ice between the shore and the victim, or it could break the victim's supporting ice. If the supporting ice breaks, the object might be the ultimate cause of drowning for any victim who holds on to it or becomes entangled. Metal objects cause problems because they're usually nonbuoyant as well as highly heat conductive. A victim who has minimal hand strength because of the cold may find his hands completely useless after grasping a metal object. Metal will steal heat from victims and rescuers alike.

A number of items typically available at a surface ice rescue make effective reach tools. Examples include rescue rope throw bags, inflated fire hose, backboard straps, seat belts, sheets, scarves, belts, tow rope, tree limbs, bench planks, wooden poles, oars, floating backboards, medical antishock trousers, stretcher mattresses, and CPR boards. If possible, extend the reach by tethering the chosen object to the shore with line, straps, sheets, or other available materials. If you aren't sure whether an item is sufficiently buoyant or not, do what you can to increase its buoyancy by attaching items such as inflated air splints, foam head stabilizers, or empty, sealed plastic bottles. A ladder can be made much safer as a rescue tool by attaching an inflated fire hose to either side of it. Duct tape works well to secure flotation in such instances. If possible, have buoyant reach tools at the ready during ice season. Use common sense as well. If a victim is progressing from the passive to the submerging state, don't waste time trying to create buoyancy for a reach tool—you must reach to him as soon as possible.

When using an inflated fire hose, attach two 150-foot deployment lines to one end of capped fire hose.

A cardinal rule of the reach method is that whoever extends the object must not wind up on the ice. Operationally trained and prepared rescuers are strictly shore-based personnel, and they should not set foot on the ice. It may be necessary to harness and tether Operational-level rescuers to immobile objects to ensure their safety, and they may require ice cleats to prevent them from slipping.

Inflated Fire Hose: Using carabiners, attach two 150-foot deployment lines to one end of capped fire hose. The lines should be stored in bags. Inflate the hose slowly with an SCBA cylinder until the hose is stiff. You should be able to depress the hose slightly with a finger. The supervisor of the operation assigns a right and a left tender to the attached deployment lines and coordinates the two line tenders and three or more hose tenders to send the hose out toward the victim. A good training program should teach personnel several techniques for sending the hose out straight.

Send the roped end of the hose beyond the victim, keeping hold of the other end of it onshore. If necessary, the line tenders can then pull the far end of the hose around the victim, forming a U. Tell the victim to hold on to the hose and even to climb on it, if possible. If the victim is strong enough to hold on, slowly bring the hose back toward shore.

Snow on the ice will make it more difficult to send the hose out straight, but if the body of water is less than 150 feet wide, you can throw a messenger tag line across to a rescuer on the opposite shore. Then, attach a heavier line to the messenger-line bag and have it pulled across to the other side. Attach the inflated fire hose to the heavy line and pull it across to the victim. If the body of water is wider, use a line-thrower gun.

Throw Bags: Although the method of deployment is a throw, a rescue bag is a reach tool, since it creates a direct connection between the victim and the rescuers. Also, if the victim is strong enough to hold on, it can bring him back to shore. Rescue rope throw bags are among the most effective of rescue tools for any season, not just winter ice rescue.

Throw Method

The throw method involves deploying a tool to a victim without creating a direct link to the rescuer. The tool is then used by the victim to establish buoyancy or to effect a self-rescue.

Be careful of throwing something that could injure the victim, damage the ice, or knock the victim away from his handhold. A common item of choice is a tire, but tires present problems. First, how do you get a tire to a victim? If you roll it, you could hit him and cause him to go under. Tires float high out of the water, and they're difficult for weak, hypothermic victims to grasp. Before throwing something, ask yourself whether the victim will more

Deploying the inflated fire hose.

Tell the victim to hold on to the hose, and even to climb onto it, if possible.

easily be able to hold on to it or the supportive ice. A good summer drill is to role-play victims and have them practice holding on to different types of objects in the water. Make a priority list of items that you can throw to victims for buoyancy or self-extrication.

Tethering Shore-Based Tenders

If an embankment is steep or slippery, secure the technicians to an immobile object with harnesses and tethers. When securing a tether line to an immobile anchor point, secure the tether line to a strap, and then attach the strap to the anchor. A strap will stay in place and handle abrasion better than a rope will. Again, use one or two nonlocking carabiners to attach the line to the strap to allow for rapid removal or to quickly attach an additional line to lengthen the tether.

If you use an ice piton, screw it in rather than pound it in. Pounding will fracture the ice, creating an unsecure anchor point. If you screw in the piton, the resulting friction will cause a small amount of melting, which will soon refreeze and provide extra security for a strong hold. For this reason, only purchase sharp-ended pitons that will easily screw into any type of hard ice.

Restrict the hot zone, ice, and water areas to fully-dressed Technician-level rescuers and tenders. The staging area should be in the warm zone.

STUDY QUESTIONS

1. When triaging victims still in need of rescue, which three types are at greatest risk?

2. An Operational-level rescue procedure that establishes a direct link with the victim is known as a _____.

3. It is best if a reach tool is _____ and _____.

4. When deploying inflated fire hose around a victim, the role of the line tenders is to _____.

5. Although commonly used, tires present several drawbacks when deployed as buoyancy tools for victims. Why?

6. If you use an ice piton for anchoring, _____ it into the ice rather than _____ it in.

9

The Rescue:
Technician-Level
Duties and Responses

If shore operations aren't feasible for the incident at hand, if shore operations have failed, or if the victim is passive or submerging, then the rescue must immediately move to "go" status. The basic protocols of responding Technician-level personnel are as follows:

1. Light, physically fit ice rescue technicians should dress with the help of certified tenders. Dress en route if possible, unless the scene is far from the road. Rescuers shouldn't travel long distances on land in ice rescue suits.

2. Designate a primary rescuer, backup rescuers, chief tenders, and tenders. As will be described later, each deployed tether line has a chief tender who is in charge of all movements of the line.

3. The primary rescuer begins the approach.

4. The backup rescuer follows twenty feet behind the primary rescuer, taking a different route. If the primary rescuer crosses over weak ice and falls in, the backup may still be able to reach the victim. A properly dressed, physically fit primary rescuer should be able to comfortably remain in ice water for at least twenty minutes, unlike the original victim, who may be seconds from submerging. Unless the primary rescuer is in immediate need of assistance, the backup rescuer's priority is to reach the victim and establish buoyancy for him.

THE ICE RESCUE TECHNICIAN

It is a given at any ice rescue incident that the ice is unsafe. Therefore, your lightest rescuers should be the only ones allowed onto it. They need to move low and slow. Depending on the situation, you may want to deploy your best technician as your backup, with your next strongest as the primary. If so, it'll be your best rescuer who takes up the slack if things don't go as planned. By being slightly behind and off-center from the primary rescuer, the backup has a complete picture of the scene and is therefore capable of guiding the response if necessary.

Overweight, forty-year-old veteran firefighters who aren't in the best physical condition don't belong in exposure suits out on the ice. Rescuers with high blood pressure shouldn't be performing the most strenuous portion of the rescue. Asthma, too, is a contraindication for ice and water rescue because of water aspiration, facial immersion, cold, high carbon dioxide levels, exertion, and physical and mental stress. Since calls can occur anytime during the day or night, insulin-dependent diabetics could be putting themselves at risk by venturing onto the ice without proper food and insulin intake. An ice rescue requires energy, strength, and physical stamina, even when all goes well.

Ice rescue technicians should be dressed in warm, comfortable clothing, including one or two pairs of good socks. Wool and Polartec™ make excellent materials under ice rescue suits, since they maintain warmth when wet.

Make sure the ground is clean, or put down a tarp to prevent rescuers from having debris stuck to their feet as they step into the suits. That debris can become quite uncomfortable during the rescue.

It is important to avoid fatiguing and stressing the technicians before they get to the scene. They shouldn't be the ones to carry gear, and they shouldn't trudge through the woods wearing ice rescue suits. Neoprene suits don't breathe, so if the rescuers perspire in them, they may already have become very cold by the time they're ready to perform the rescue. In-water ice rescue suits are difficult to move in, and so will unnecessarily fatigue rescuers who try to walk long distances in them. Also, wearing them on land increases the risk of puncturing them. Have support personnel and transport devices move the gear so as not to compromise the rescuers' strength, and have the rescuers dress at the scene for incidents far from the road.

When dressing, the tender, not the technician, should close the suit's zipper, making sure that no clothing is in the way. The tech-

Never tether a technician solely to his back harness point, since pulling him out of a hole this way would put undue strain on his spine.

nicians should then don ankle weights, if available, as well as the harness. Secure one end of the tether line to the harness and the other end to an immobile object. Then, put the rescue flotation sling up over the technician's dominant arm and over his head so that it lies diagonally across his torso. Slings will be covered in greater detail below. How the harness is tethered depends on how the rescue will be performed.

No matter what procedure you use, never tether a technician solely to his back harness point. Extricating him from a hole this way would prove difficult, putting undue strain on his body as he is yanked up onto the ice shelf, stressing his spine the wrong way. If the rescuer is tethered slightly off-center by the front of the harness, then the tenders can easily assist him out of the hole. Tension on the front tether point will put the technician's body in the low, horizontal, face-down position needed for extrication. If the front harness tether point is lower than the solar plexus region, it may restrict breathing and will force the wearer into a vertical position. For that reason, rappelling harnesses don't work well for water or ice rescue.

If no ice transport device is available, the following technique may prove useful for tethering a technician. Using a figure 8 loop, attach the end of the tether line to a locking carabiner on the back ring of the harness. Place a piece of duct tape over the carabiner lock to prevent it from freezing shut, to prevent the line from abrading the lock, and to prevent the line from unscrewing the lock. Then, ask the technician to raise one arm out to his side.

The technician frees the second loop from his harness and attaches it to the victim's sling.

Make a second, hand-sized loop in the line at the technician's hand. Attach this second loop to the technician's front harness tether point with a nonlocking carabiner or, preferably, a spinnaker shackle or other sturdy quick-release system. This second loop attachment allows tension to come from the front part of the harness, allowing tenders to assist the technician out of the ice hole, if necessary.

Later, after the technician has secured a rescue flotation sling on the victim, the technician can release the spinnaker shackle or nonlocking carabiner to free the second loop from his harness so that it can be attached to the sling. Then, when the technician tells the shore personnel to take up the slack, the victim will be assisted out of the hole. Once the victim is out of the hole, the technician can lie alongside him, and both can be pulled to shore by the tenders. The victim's sling is tethered to the technician's previous front tether point, and the technician is still tethered to the back harness point. If the technician falls through a hole during egress, he can disconnect the second loop from the victim's

Once the victim is out of the hole, the technician can lie alongside him, and both can be pulled to shore by the tenders.

sling, reattach it to the front harness tether point, and allow tenders to assist with the extrication. Once out of the hole, the second loop is disconnected from the harness and resecured to the victim's sling.

When no victim transport device is available, this procedure allows a technician to lie on his back and put a small child on his stomach, facing up. The rescuer's body becomes the transport device. When the shore personnel pull, both the slinged child and the rescuer will be pulled together. Technicians shouldn't use this technique with a heavier person, since the potential for injury to the rescuer is too great. The rescuer's chest can be crushed, or his head or spine could be damaged by hitting a piece of ice or debris. For reasons of everyone's safety, you should do whatever possible to obtain an ice board, sled, or kayak.

One common question is, "Why not have the technician bring a second tether line to attach to the victim's sling?" One reason is that the two lines will most likely intertwine and effectively become one, essentially wasting the second line and hindering the operation. The intertwining of double lines makes it very difficult for shore personnel to control the lines effectively, and the rescuer cannot disconnect quickly from the victim if necessary.

Another procedure once commonly taught is to have the rescuer wrap a line around the victim's torso, thereby securing the victim to the rescuer. This procedure is no longer recommended. The rescuer won't be able to disconnect quickly from the victim in the event of a mishap, and this could result in injury to either party. Also, if the victim were to panic, the rescuer could be injured. Typically, the victim winds up on top of the rescuer, which also endangers the rescuer. This procedure makes it difficult to get the victim out of the hole, and once the rescuer and victim are attached to each other, they will end up stuck together in an ice hole if they puncture the cap. If the rescuer disconnects from the victim to get out of the hole, the victim no longer has buoyancy or a tether and may quickly submerge.

There are additional methods of tethering the technician so that the rescuer ends up with a front tether point, once the victim's sling is secured to the second loop. Although these methods are very effective, they are more complicated and require redundant training and drills on the part of both tenders and technicians.

THE CHIEF TENDER

The two players who physically perform the rescue are the ice technician and his chief tender. These two must work with practiced precision. A poorly trained tender can not only hinder an operation, he can also cause injury to both the technician and the victim. Tenders must be able to give and take just the right amount of slack, and they need to be able to stop the line to prevent injury. Since it is common for there to be two to four tenders on a technician's tether line to manage a long line and perhaps a heavy or inefficient transport device, it is imperative that the tenders act as one entity. The only way such coordination can occur is if one tender is in charge of all the tenders on the line.

Just as a nozzleman controls a fire hose, the chief tender controls a rescuer's deployment tether line. The tender closest to the water is designated the chief tender, and he controls all line movement. Since there are at least two rescuers, a primary and a backup, each on a separate line, there will be at least two chief tenders.

Another important function of tenders is to make certain that the technician is checked out before dressing, is dressed properly without becoming fatigued, and is attended to once he's off the ice. It is the chief tender's responsibility to ensure that the technician follows all procedures before and after the operation. The chief tender never assists the victim. His full responsibility is to his technician until that technician has dressed down and has been checked out.

A chief tender provides supervision to maintain a manageable span of control. He calls aloud all signals received from the rescuer and gives the command to carry out the signal. This provides organization and lets the officer in charge know what is happening between the rescuer and the shore personnel. If the rescuers are difficult to see because they're so far away, the chief tender might hold binoculars rather than the line. Other tenders can work the line. The chief tender must be able to receive signals and give commands.

Slow, controlled line pulling is critical for a safe operation. A chief tender ensures and controls a maximum line-pull speed of one foot per second. If a rescuer or transport device punctures the ice, someone may be injured if he is pulled too hard into the ice roof, and he may even be pulled under. A person may also sustain injury if he is pulled into conflict with a rock or some other hazard frozen into the ice. The one-foot-per-second rule allows the person or object to come to a quick, full stop the moment the rescuers stop pulling. You can practice pulling at this rate by reaching one foot up on the line with each successive pull, while counting off the seconds aloud.

The chief tender shouldn't allow the line tenders to run up the shore with the line, since it is far more difficult to control the speed of the line when using this technique. Also, it's too easy for tenders to slip and fall, especially on snowy or icy ground. A running tender has to look where he is going, taking his attention away from the technician and the chief tender. Finally, running tenders will exhaust themselves more quickly than standing tenders, and they're more likely to perspire, resulting in evaporative heat loss and subsequent cold stress.

Chief tenders should make sure that all of the other tenders on the line are wearing proper gloves, personal flotation devices, and other equipment, and they must ensure that all of them know what signals will be used. A chief tender himself should wear antislip shoes and may need to be tethered with a harness to prevent him from slipping or falling.

TECHNICIAN-LEVEL RESCUE PROCEDURES

The rescuers' approach should not destroy any additional ice, if possible. It's important to use the direct line between the hole and the shore only for egress. If the rescuers destroy the ice along this path, then they must bring the victim to shore by way of a less direct route, since transporting a victim through broken ice and open water isn't a good choice. Also, if the rescuers destroy the supportive ice in front of the victim, there's a greater chance that he will submerge. Therefore, save the most direct route for taking the victim to shore. The rescuers' best action is to follow a route to the victim that is slightly off the centerline.

Rescuers can approach the victim using several different procedures, depending on the conditions of the ice. Whatever method they use, the rescuers should wear a flotation sling to provide buoyancy for the victim on contact.

Crouch and Knee Shuffle With Ice Pole

If the ice appears to hold your weight, crouch low and slowly shuffle or walk, on your knees or feet, using an ice pole to test the ice in front of you. If you choose to walk, you should wear ice cleats to prevent yourself from slipping and falling.

Crawling and Prone Movement

If the ice is too weak for the above methods, try crawling on your hands and knees to distribute your weight over a larger surface area.

If the ice is too weak for crawling, then you should lie flat on your stomach with your arms and legs outstretched. You can then proceed in a prone position, slithering like a snake or an alligator. Be careful not to push your elbows through the ice, which can result in facial immersion. A good pair of ice awls makes this method of approach much faster and more efficient, since they allow you to pull yourself across the ice without putting any weight on your knees or elbows.

Approaching With Sleds, Boards, or Kayaks

A good sled, board, or light kayak should be relatively easy to pull across ice and even snow. You may crouch and crawl as described above while pulling the transport tool behind you. If the ice is weak, you should be able to lie on the transport tool and pull yourself along with ice awls. Make sure that the awls float in case you encounter a stretch of open water.

Approaching With Boats or Inflatable Ramps

As mentioned in Chapter 6, using boats and ramps for approaches can create a number of potential problems. The issue of breaking up the ice to move boats is an important one. If the supportive ice is broken up, the victim is put at much greater risk of submersion by the rescuers themselves, possibly leaving them unable to reach the victim quickly enough before the submergence occurs. If the victim goes down, his location might be lost. If there is no profile map of the location, the destruction of the hole means that the rescuers and the arriving dive team will have to waste precious time searching for him. Also, if the ice is broken up, the responding dive team will be forced to operate solely from their boats, since they'll no longer be able to use the ice as a platform.

The primary and backup technicians should reestablish ASAPS status when they're approximately ten to fifteen feet away from their target. This will determine which victim to approach first, what techniques to use, how quickly the technicians must move, and whether or not the dive team should be fully dressed and ready to go.

Rescue Flotation Sling

Buoyancy for the victim should be completed within ten seconds of physical contact. One of the best methods involves using the rescue flotation sling, which prevents submergence and helps rescuers remove the victim from the hole. Rescue flotation slings can be used all year round in a variety of water rescue situations.

If the victim can't get out of the hole without assistance, use the following procedures. Wear the sling over your shoulder, across your chest and back, and down to your side on the same side as your dominant arm. For example, if you are right-handed, wear the sling over your left shoulder, around your body, and down to your right side. Worn in this manner, the sling won't affect the approach procedure.

When you approach the victim, approach him from his side. That way, you'll be less apt to destroy the ice that's supporting him. Lie flat on your belly, with your legs outstretched and wide, so that your body is perpendicular to his. This position will create the greatest drag for you to anchor yourself in place. If you face the victim in a position that's parallel to him, he may be able to pull you into the hole as soon as he grabs you or your pole. Do not put yourself within reach of the victim until you are securely in posi-

When worn correctly over your shoulder, the rescue sling won't interfere with the approach procedure.

tion. Then, give the shore personnel a signal to stop so that they will neither give nor take any line until you give the command. This will make you an even stronger anchor.

Next, using your dominant hand, firmly grasp the same-hand wrist on your victim. (If you are right-handed, grab his right wrist with your right hand.) The wrist provides a better grip than the hand. The right-to-right or left-to-left configuration puts your arms across the victim's body in a protective position so that if he suddenly becomes aggressive and attempts to climb out on top of your head, you can quickly bring up your elbow to protect yourself. By grabbing his left wrist with your right hand, you wouldn't be able to perform this defensive maneuver. Allowing the victim to crawl out over your body is certainly a viable rescue procedure, but you need to be in control of it to prevent your face from being smashed against the ice, or perhaps concentrating too much weight on a weak surface.

In one smooth motion, slide the sling up over your head, down your arm, and up the victim's arm, catching the victim's armpit with the sling. Once part of the sling is under the victim's armpit, it's already providing some buoyancy for him. Then, slide the rest of the sling over the victim's head, and bring his other arm through the sling so that it sits under both armpits. This latter maneuver can be performed in two ways. If the victim is able to help, ask him either to put his own arm through the sling or to lean toward you,

In one smooth motion, slide the sling down your arm, over the victim's head, and under his armpit.

Never let go of the victim until the sling is fully secured to him.

Snug the sling fully against the victim's chest.

allowing you to scoop the arm into the sling. If the victim is passive or submerging, release the sling and grab his clothing in the chest area with your nondominant hand. Then, use your dominant hand to reach inside the sling, grab the victim's other arm, and pull it through the sling. Never let go of the victim until the sling is fully secured to him.

Hold the sling, and secure the self-fasteners around it to snug it up against the victim's chest. This procedure will prevent him from sliding out of it should he become unconscious. Attach the sling either to the released second loop of the rescuer's tether line or to the transport device so that the shore tenders can assist the victim out of the hole.

If the victim is aggressive, anchor yourself with the length of your body perpendicular to the victim, and tell the shore personnel to hold you securely in place. Extend the sling to the victim and instruct him to climb up it. Once the victim is out of the hole, have him roll away to a safer area, then sling him as described above.

The Rear Approach

To make a successful rear approach to sling a victim, rescuers must train and drill frequently, since simple wrong maneuvers can easily provoke submergence. The only time to attempt such an approach is when the ice to the victim's front and side is so weak that moving across it would compromise his support. Maintaining verbal communication is imperative during a rear approach, including telling the victim not to turn around.

By this method, the rescuer enters the hole, approaches the victim from the rear, and supports him with one arm wrapped around his side and chest. He uses his dominant arm to extend the sling over the victim's hand and arm (if right-handed, slip the sling over the victim's right hand), then he scoops them up into the sling.

Once the sling is successfully up the victim's arm and supporting the armpit, the rescuer gently places the sling over the victim's head, then down to scoop up the other arm. The rescuer must avoid pushing down on the victim's head or shoulder, which could cause submergence. Once the sling is in place, the self-fasteners must be secured.

The sling must now be turned 180 degrees to bring the tether point to the front of the victim for extrication. After this is done, the rescuer exits the hole and follows the standard procedure for tethering and extricating a victim in a sling. If the supportive ice breaks at this point, it isn't such a problem, since the victim is now buoyant and no longer needs the ice to stay afloat.

Several problems with the rear approach must be addressed during training and in consideration of SOGs. First, approaching from the rear makes it too easy to pull the victim away from supportive ice or to push him under before he is properly secured. A rescuer's tendency is to grab on to something for stability. In a rear approach, the victim will be that object. If the victim is pulled off of his supportive ice, it will take a significant amount of training and skill to sling him before panic or submergence ensues. Also, you should never use the rear approach with aggressive victims. Since this method puts you in the water and at his level, an aggressive victim is apt to turn suddenly and grab you once he becomes aware that you're there. This is inherently dangerous from the standpoint of proper lifesaving techniques. Once the victim has been slinged, the rescuer must get out of the hole to extricate the victim, and he must do so without letting go of the sling, which is quite difficult. The rescuer wouldn't be able to lift the victim out while they were both still in the water, since pushing up on the victim would only drive the rescuer under. Front and side methods continually pull the victim toward the supportive ice, but working from the rear has a tendency to pull him away from it. Finally, working from the front and side affords eye-to-eye contact and face-to-face communication, which can enhance the rescue process.

The Ice Pole Hand-Loop Grab

If the supporting ice is very weak, you don't want to risk breaking it up before having a secure grip on the victim. For the rescuer, the rear approach is one answer to this problem, but the ice pole hand-loop grab is a better one.

To perform the ice pole hand-loop grab, anchor yourself in the sideways, prone position described earlier, slightly more than a pole's length away from the victim and slightly to the victim's side. Maintain appropriate verbal communication. Extend the end of the pole and snare the victim's hand (or better yet, his wrist) with the loop. Gently twist the pole, entrapping the hand or wrist securely. The goal is that, if the victim submerges before you can reach him, you can at least pull him back up as soon as you're able to reach him to grab his hand and arm.

Once the victim's hand is trapped in the loop, ask your tenders for slack, then slowly move toward the victim without pulling on the pole. Once you have a good grasp on the victim's wrist with your dominant hand, release the loop and move the pole aside. Finally, sling the victim as described above.

If the rescuer and victim both end up in open water, and if the victim is passive or submerging, pull him onto your body to establish buoyancy for him. Maintain strong verbal communication, encouraging him to stay alert and assist in his own rescue. Securely hold the victim with one arm and your legs. With the other arm, pull the sling slowly over your head and bring the sling up over the victim's arm and head. Pull the victim's other arm through and secure the self-fasteners.

This technique requires training and much practice with mock victims who are adept at playing unconscious victims.

Extricating the Victim From the Hole

Whatever type of transport device or procedure is used, the victim must be extricated safely, gently, horizontally, and with as little compromise to the airway as possible. If the only transport device

When performing the ice pole hand-loop grab, it's best to snare the victim by the wrist.

available is nonbuoyant, remember this rule: Never strap a victim to a nonbuoyant device.

Protect yourself from aggressive victims. Communicate with them and calm them down. As trained lifeguards will tell you, don't get within arm's reach of an aggressive victim. Rather, pass off the transport device. An aggressive victim or one who is capable of self-rescue may be able to climb up onto the device by himself, providing that the device is designed for it. A good ice rescue board, for example, is ringed with a flotation collar that is easy to grab and that has several handhold straps. Once the victim is on the board and calmed down, the rescuer secures the sling to him and the board. The rescuer then assesses the victim and prepares to move to the side of the victim's hole before advancing further. This will decrease the chances of breaking the supporting ice.

If the victim is capable of self-rescue, the rescuer can use the proper techniques to extend a pole and instruct him to slowly kick his feet. The victim then climbs up the pole to the rescuer, who secures the sling to him. If the rescuer determines that the ice is too weak to approach the victim any closer than a pole length, or if the victim is close to submerging, then the rescuer can snare the

118 victim's hand or wrist with the strap on the pole. The rescuer can then approach and sling him to establish buoyancy.

Alert and passive victims may or may not be able to follow simple commands, but they probably lack the strength to assist in their own extrication. These victims are the most likely to submerge if they lose hold of supportive ice. Therefore, establishing buoyancy on contact is imperative.

When using ice boards, ice-rigged flotation backboards, or sleds that can be partially submerged by the rescuer, establish buoyancy for the victim with a rescue flotation sling. Submerge one end of the transport device, then gently slide the victim onto it.

In the case of hard kayaks and sleds that cannot be partially submerged by the rescuer, try to use the basic board technique described above. The main difference will be that the rescuer won't be able to push the stern of the device underwater to slide the victim onto it. The rescuers may have to reach over and pull the victim on board, or they may have to use other tools, such as roll-up straps. Kayak rolls can also be used to get a victim into a kayak.

If boats, ramps, or inflatable kayaks are used, establish buoyancy for the victim with a rescue flotation sling, as described earlier, and use roll-up straps or netting to roll the victim gently, horizontally into the vessel.

If the victim is submerging, establish buoyancy for him immediately. If you can grab the victim from the transport device, do so, but be very careful of hitting him with the device or disturbing the water enough to bring about his final submergence. If the victim cannot be safely reached from the transport device, then the rescue moves to live-bait status, in which a technician enters the water to reach him. Again, be careful not to disturb the water too much. Immediately establish buoyancy for the victim with a rescue flotation sling, as described earlier. You may also use a reach pole with a shepherd's crook to pull a submerging victim closer to the transport device.

Recovering Submerged Victims

A reach pole may also be used in the case of a submerged victim, provided that the water is shallow enough and doing so won't hinder a dive team's ongoing operation. If the dive team is still en route, use a reach pole. Add enough extensions so that it is long enough to reach the bottom while allowing your hands to be comfortably out of the water. Make sure your hands are protected with warm gloves that won't lose their insulation when wet. Sweep

An aggressive victim or one who is capable of self-rescue may be able to climb onto the transport device by himself.

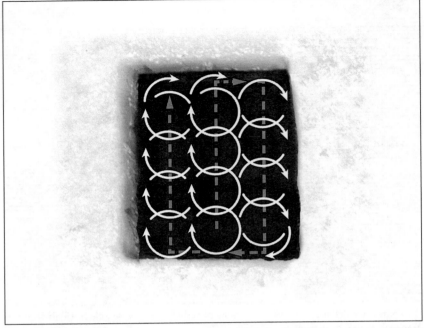

When searching with a reach pole for a submerged victim, sweep the bottom using circular motions two to three feet in diameter.

using circular motions two to three feet in diameter. If you find nothing there, shift your sweep to the side, overlapping the circular motions you make. Work systematically so that you fully cover a given area. Search the largest amount of area with the least amount of effort, and check much of that area twice. Circular patterns won't tend to push the search object out of the way as a back-and-forth sweep might. If the water is relatively clear and shallow, attaching a small waterproof light to the pole shaft just above the grapnel extension may even allow you to see the victim.

Rescuing Pets

Use the shepherd's crook of a reach pole to pull in the hind end of the animal, to grab on to its collar, or to pass a sling lasso around its body. Remember, though, that no animal is worth risking a rescuer's life.

Ice Pole

An ice pole can be used to approach and extricate an aggressive victim or one who is capable of self-rescue. When you are

When extending an ice pole to a victim, lie perpendicular to the victim and have the shore tenders taughten the line.

within ten feet of the victim, tell him to hang on and to follow your instructions. Lie on your stomach, perpendicular to the victim, with your legs spread wide. Push the looped end of the pole out toward the victim, then shuffle yourself sideways so that the victim is within inches of reaching the pole. Then, give the shore personnel the signal to tension the line. Push the end of the pole to the victim and command him to grab the pole, kick with his feet, and stay low as he climbs up the pole. Once the victim is out of the hole, you can shuffle your body backward to help get him clear of the hole. The personnel on shore should feel the slack and know to take it up. Tell the victim to slowly and gently roll away from the hole, toward you. Establish buoyancy for him and transport him to shore.

If the ice is so weak that you're afraid it will break under your weight, then use the pole to make the first contact to secure the victim and prevent submergence. Once the shore personnel are holding the line tight, command the victim to hold on to the ice. Reach the looped end of the pole close enough to one of his hands. Preferably, it should be the hand opposite the one that you'll use when putting on the sling. As described earlier, once the loop is over the hand or wrist, twist the pole to secure it. You now have the victim secured in such a way that, if he does submerge, you have his hand and can pull him to the surface. Work your way to the victim, but do not pull on the pole. Nor should you attempt to pull the victim out by the pole handhold. Once you reach his wrist, make contact and establish buoyancy with the rescue flotation sling. Then, release the pole from the victim's wrist.

THE BACKUP ICE RESCUE TECHNICIAN

The primary rescuer is first on the ice, moving low and slow off to one side of the direct line toward the victim. The backup rescuer should be approximately twenty feet behind him, coming from another angle. Think of ice as being disposable—you only want to use it once. A backup ice rescue technician isn't necessarily a safety technician for the primary technician; rather, he functions as a backup to reach the victim if the primary technician is delayed by falling through the ice. This role is one of the reasons the backup technician approaches the victim along a different route. It the primary technician falls through, he can stay in the water safely for a far longer time than the original victim, who probably isn't wearing an insulated, full-flotation dry suit. Also, the primary technician should be able to extricate himself with his ice tools, the assistance of his tenders, and the knowledge he has gained through training.

It could happen, though, that the primary rescuer finds himself in need of immediate assistance. For example, suppose he banged his head when puncturing the ice and is now bleeding from a head injury. If such is the case, the backup rescuer may be directed to aid the primary technician. The backup is twenty feet behind and to the side of the primary, so he can travel across virgin ice to reach him. This decreases the risk of puncturing the ice. Also, by being behind the primary, the backup always has the primary technician in view, and therefore has the best vantage point in case anything goes wrong. The primary technician may be too far from his tenders for them to see whether he requires immediate assistance.

If the primary rescuer falls through the ice and is okay, the backup rescuer, who is approaching from a different angle, should bypass the primary rescuer and continue toward the victim to establish buoyancy for him as quickly as possible. When the primary rescuer is out of the hole, he can bring the transport device to the victim.

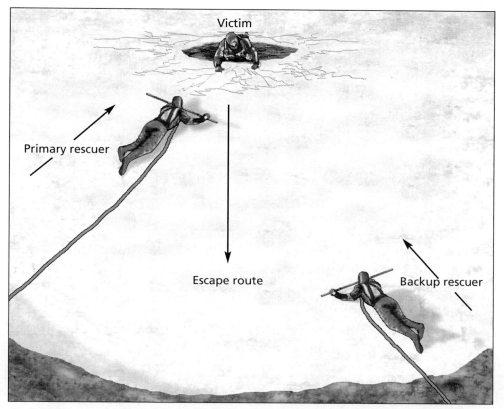

Victim

Primary rescuer

Escape route

Backup rescuer

The backup rescuer should follow approximately twenty feet behind the primary rescuer, taking a different route to the victim.

Of course, the backup technician should carry a rescue flotation sling and ice awls in the same manner as the primary technician. He may also want to carry a rescue rope throw bag clipped to his harness, which can be used to aid the primary technician, provide an immediate lifeline to a victim if the primary is delayed, and assist in a multiple-victim situation.

When the rescue is taking place hundreds of feet from shore, you may require a second backup and on-ice support personnel. The angle of approach for the third backup should be different again. This is to give each rescuer the best access possible, on new and undamaged ice. Every time you move across ice, you take the chance of damaging or weakening it. Also, taking a third, independent route means that you'll have one more chance of finding ice that can support a rescuer.

STUDY QUESTIONS

1. True or false: Ice rescue technicians should dress en route, if possible, unless the site of the incident is far from the road.

2. Once deployed on the ice, an ice rescue technician should move _____ and _____.

3. Taking a different route, the backup rescuer should follow approximately _____ feet behind the primary rescuer.

4. Does the chief tender ever assist the victim?

5. The chief tender should ensure a maximum line-pull speed of _____.

6. The direct line between the hole and the shoreline should be used only for _____.

7. Name three drawbacks to breaking up ice so as to move a boat to the victim.

8. Buoyancy for the victim should be completed within _____ of physical contact.

9. Is it best to approach the victim from the side, the front, or behind?

10. True or false: With aggressive victims, it is best to use the rear approach.

11. Although called the ice pole hand-loop grab, it is best to use the loop to snare the victim's _____.

12. When searching with a reach pole for a submerged victim in shallow water, you should make _____ motions. How large should they be?

10

Communication and Line Tending

Just as verbal communication is so important in our daily lives, so is the ability to communicate during any rescue. Operations on ice are no different. The on-ice rescuer must be able to communicate with tenders to tell them when to pull, when to stop, and when to give more slack to the line. This communication can be accomplished with a series of hand, line, or whistle signals. Without this communication, tenders won't be able to provide assistance to the rescuers, and they may even hinder the rescue effort.

Suppose, for example, that the rescuer is close to the hole. He assesses the victim as being aggressive but capable of self-rescue. While talking to him, he lies down near the side of the hole, positioning himself to hand the victim his ice pole so that the victim can climb up. He lies down far enough so that the victim cannot grab the pole before he is positioned properly. Without a communication system, the line tenders mistakenly think that the technician needs more slack in the line so he can get closer to the victim. They give him slack, the victim grabs the pole, and the technician is pulled into the hole. A simple communication system could prevent just such an incident from occurring.

There are three basic commands for the shore personnel to follow: tension the line and slowly pull; slacken the line; and stop. *Stop* means "do not give or take line."

The Stop command given as an arm straight up with a fist.

STOP SIGNALS

Arm signal:	Arm straight up with a fist.
Light signal:	Hold light straight up toward shore, or wave horizontally from left to right.
Whistle signal:	One blast.
Line-pull signal:	One pull.

Of the three basic signals, the *Stop* command will cause the fewest problems if the tenders mistakenly perform it when the rescuer asks for something else. Suppose the rescuer gives the signal for more slack because he wants to get closer to the victim, but the tenders misread it as *Stop*. The only result of this mistake is that nothing happens. The rescuer just stays where he is. On the other hand, if the tenders misread the communication as *Tension the line and slowly pull,* then the rescuer will end up farther from the victim, which could result in the victim submerging.

If the rescuer requests *Tension the line and slowly pull* to assist the slinged victim out of the hole but instead receives *Stop*, all that will happen is that the victim stays in the hole until the signal is corrected. If, on the other hand, the tenders misread it and give the rescuer more line, the rescuer could end up in the hole with an aggressive victim. Therefore, the commands should be designed so that, if any signal is misread, the tenders will think the rescuer is giving the *Stop* command.

SLACKEN THE LINE

Slacken the line means to issue three to five feet of slack each time the rescuer gives the signal.

Arm and light signals	Wave arm up and down vertically.
Whistle signal:	Two blasts.
Line-pull signal:	Two pulls.

TENSION THE LINE AND SLOWLY PULL

Tension the line and slowly pull means to pull the line gently and continuously at a maximum speed of one foot per second. Stop pulling when the stop signal is given.

Arm and light signals:	Arm up, making large circular motions.
Whistle signal:	Three blasts.
Line-pull signal:	Three pulls.

Tension the line and slowly pull is the signal to assist the victim or rescuer out of a hole and to bring him to shore. It is important to pull slowly and continuously. If the line is pulled with stop-and-go motions, the rescuer's suit or tools could have time to freeze to the ice. Also, stop-and-go motions are more uncomfortable for the person being pulled. If the rescuer or victim is pulled too quickly, he could be pulled underwater, under the ice roof, into an obstacle, or tipped off the transport device. The rescuer needs time to control every aspect of the situation.

HELP

Help means to send the backup rescuer to the primary rescuer or to where he is pointing.

Arm and light signals:	Arm up, sweeping side-to-side over the head.
Whistle signal:	Four blasts, with possible continuous four-blast repetition.
Line-pull signal:	Four pulls, with possible continuous four-pull repetition.

This signal is for one of those rare situations where the primary rescuer may send the backup directly to the victim if the primary rescuer cannot continue forward. If the original victim is passive or submerging, then immediately establishing buoyancy for him is critical. As long as the primary isn't experiencing some other problem, such as respiratory distress or trauma, then the original victim should be attended to first.

Still, the chief tender of the primary rescuer has the ultimate say as to where the backup rescuer will go, even if the primary gives the help signal and points to the victim. If the chief tender decides that the primary rescuer is in real distress and appears to need immediate assistance in a manner that cannot be achieved by shore personnel, then he will send the backup rescuer to help the primary rescuer first. Rescuer well-being is always the first priority.

In the case of multiple victims, the primary rescuer might give the help signal and then point to one of the victims. The primary rescuer can then approach the neediest victim first while the backup rescuer progresses toward a second victim.

Depending on the distance from shore, the amount of snow on the ice, and the distance of open water between shore and the rescuer, line pulls and whistle signals are the best options. Light signals and long-distance night operations require additional training and are beyond the scope of this book.

Sometimes both of the primary rescuer's hands are occupied, and thus he cannot give a signal. The backup rescuer can relay communication from the primary to the line tenders. If the backup is in the proper position twenty feet from the primary, then he should be able to hear the primary's shouts to stop, slacken, or pull. The backup can simply point to the primary with one hand and give the signal with the other, telling the tenders that the signals apply to the primary rescuer.

It's helpful to put the signals on ID-size laminated cards and attach them to every PFD to help new tenders remember them. Hang the cards upside down so the tenders can read them without letting go of the line.

130 STUDY QUESTIONS

1. The three basic signals for shore personnel to follow are _____, _____, and _____.

2. Which of the three basic commands will cause the fewest problems if the tenders mistakenly perform it when the rescuer asks for something else?

3. What is the arm signal for *Stop*?

4. What is the line-pull signal for *Stop*?

5. What is the arm signal for *Slacken the line*?

6. What is the whistle signal for *Slacken the line*?

7. What is the arm signal for *Tension the line and slowly pull*?

8. What is the whistle signal for *Tension the line and slowly pull*?

9. What are the arm and light signals for *Help*?

10. What is the line-pull signal for *Help*?

11

Special Ice Rescue Situations

VEHICLES

Rescues involving vehicles on the ice or submerged beneath it are among the most complicated, dangerous situations that rescuers can encounter. A vehicle can submerge at any time, and if you become entangled with it, you will go down as well. Your movements alone could cause it to go under. Also, if you submerge your head to look or reach inside the vehicle and you become entangled, you could drown. To add to the hazards, sealed containers with air or gas inside (baby bottles, spare tires, soda bottles, and the like) become upward-moving projectiles if they are submerged and then freed. Some fuels can cause severe burns if they contact the skin directly, and the aspiration of water contaminated by fuels can result in lipoid pneumonia.

For incidents involving vehicles and water, wear at least two pairs of shears, or one pair of shears and one safety belt cutter, on your harness. Do not submerge your head into the vehicle if you aren't a certified ice diving rescue technician using scuba and trained in underwater vehicle extrication. Do not get your face wet with the contaminated water. Beware of buoyant projectiles.

If the vehicle is about to submerge further, get clear of it. If it's still on the surface, have the occupants gently open all of the doors, if possible. Open doors act as big wedges, reducing the speed at which the vehicle penetrates ice that has given way. If you can't open the doors, have the occupants open the windows, allowing an escape route in case the vehicle submerges.

131

Vehicles that are submerged below door level won't allow for this advantage. Open doors won't reduce the speed at which the vehicle descends after it has passed below the surface.

Try to verify that the vehicle isn't being held at the surface by air inside of it. If it is, any small movement will force the air out and the vehicle will submerge. If such is the case, stay away from the vehicle and make it a dive operation. The SOGs may require that the dive team be trained in underwater extrication and equipped with dry suits, full-face masks, and a hardwired communication system whether the vehicle is submerged or on the surface.

If ice rescue technicians might have to work with a vehicle in the water, they should be prepared with silicone dive masks to protect their eyes from leaking fuel. If rescuers can no longer see because fuel is burning their eyes, they not only become useless in the rescue effort, they may also become victims themselves.

Victims must move quickly but gently to prevent further damage to the ice and the hole. Allow as few rescuers as possible to make contact with the vehicle. Securing lines or cables to it can be dangerous at best unless you understand distances and loading angles. At minimum, lines can trap, snare, or entangle rescue personnel.

SNOWMOBILES

Wherever you find snow and ice, you will likely find snowmobilers.

Participants of this sport may unknowingly end up on a snow-covered, frozen lake. They may participate in contests to see who can jump over the longest stretch of open water. They may joyously ride repeatedly in the same circular pattern until the ice melts and cracks. They sometimes (and in some areas, *often*) do these things in remote locations, far away from roads and easy access points. Alcohol is frequently involved.

Snowmobilers are often found farther out from shore than skaters and walkers. They are more likely to have compounding trauma problems, and their helmets and heavy boots can complicate matters. Moreover, fuel from the machine can pose a hazard to the victim, his rescuers, and the environment.

During such responses, the load on the rescuers can be greatly increased. EMS personnel and equipment must often be

A snowmobiler who has been involved in an accident may require stabilization of the head and neck.

transported to remote locations. Equipment to handle injuries may have to be transported across the ice. Spinal immobilization units, backboards, or rescue sleds must float. Inflatable units on the market are specifically designed for this type of extended-range rescue. Some sleds, for example, can easily be pulled by a rescuer's snowmobile and make excellent tools for long-distance operations. Side tubes allow the victim to be more sheltered for long treks to safety than he would be if he were on an ice board or a flat sled.

An important caution! Once a patient has been strapped to a transport device, pulling the device with a snowmobile or other vehicle is not recommended. If the vehicle breaks through the ice, the victim may be injured or killed. If you use a vehicle because the distance to be traveled is considerable, make sure to drive it cautiously.

A snowmobiler who has been involved in an accident may require head and neck stabilization. A good ice rescue board is one that serves as a very effective backboard and transport device. Two built-in straps secure the victim's torso. Optional plastic stick-on head and neck stabilizers can also prove useful.

DRIFT ICE

Common sense tells us what the potential damage can be if rescuers find themselves caught between two or more pieces of drift ice. The human body is fragile and can easily succumb to two pieces of colliding ice.

If you find yourself caught in such a situation, move slowly and think. Do not push the free-floating pieces of ice with any great force, since action creates reaction. If you push away a block of ice and it hits another, there is a good chance that it will rebound toward you. When ice moves, more ice will move, and if you're between the pieces, you'll become part of the action. If the pieces are small, move them slowly and watch how they react with other pieces of ice near them. Larger pieces can often be used to your advantage. If possible, use ice awls to gently climb up on them, letting them become the buffer between you and other pieces of moving ice. Keep your arms and legs out of the way.

ICE CANOPY

In some cases, the water level may drop below an already-frozen ice cap, leaving a space between the ice and the water's surface. This situation occurs particularly with reservoirs, which freeze over and can then be partially drained. Ski ponds, too, may freeze on top and then be drained for snowmaking purposes. The level of the water may be anything from a few inches to a few feet beneath the ice. In such situations, the ice has no water support beneath it and it breaks easily. To further complicate matters, once a victim breaks through the ice, he cannot easily be seen from the shore and may even drift under the still-intact ice roof. The combination of weak ice and an unseen victim makes this rescue one of the most difficult and dangerous imaginable.

Most importantly, no matter what your level of training, *do not go onto the ice*. Any rescuer going onto the ice will fall through, since there is no water beneath it to support the rescuer's weight. Furthermore, the rescuer will then be below the lip of the ice, making self-extrication very difficult. In addition, when the rescuer falls through, he may be injured by large chunks of ice, and penetrating the canopy this way can bring huge ice chunks crashing down on the victim as well.

Surface rescue in this situation is simply too dangerous for both victims and rescuers. Instead, this operation should be handled by divers wearing helmets. The dive team should cut a hole as close to shore as possible and then deploy under the ice, although remaining at the surface. On reaching the victim, they should establish buoyancy for him and return him to the original hole—once again, under the ice roof.

Even with divers, this type of rescue is extremely dangerous to rescuers and victims because of the constant threat of roof collapse. If the dive coordinator decides that the victim should be given one of the diver's pony regulators for the return to shore, then the victim should also be given a mask to allow him to breathe more easily should he suddenly be submerged in the frigid water due to collapse of the ice.

Although this book addresses surface rescue, it should be noted that the regulators used by divers for this type of rescue must be environmentally sealed. Using a regulator at the surface in ice water creates the highest possibility of freeze-up possible. Be sure divers are trained and properly equipped for any rescue under an ice roof.

MOVING WATER

An ice rescue performed in moving water is a very advanced form of rescue. Divers with only ice-diving rescue certification or, worse, sport ice-diving certification should not perform a moving-water ice operation. Sadly, most dive teams aren't even trained enough to know how to calculate water speed to know the limits of tethered diving in general, much less their own.

The most common method of calculating water speed is in knots. One knot equals a movement of one hundred feet per minute.

If the water is moving faster than one-half knot (fifty feet per minute, or twenty-five feet in thirty seconds), then tethered divers must work off a platform and can no longer work from the shore. Divers won't be able to maintain a good pattern if forced to work with or against the current. Working cross-current is better. Anything over one knot is considered fast-water diving and requires special procedures. For anything over two-and-a-half knots, divers can no longer be tethered, which means they should not be diving in conditions of low visibility, and they absolutely cannot dive under an overhead environment, such as an ice roof.

Where will a body be in moving water? As mentioned in Chapter 7, once a corpse is on the bottom, it will stay there until body gases are produced, causing it to gently drift while it ascends and descends. When it reaches shallow water and enough gases have been produced, the body will surface. In cold water, this process could take weeks or months. In cold, deep water, the body may never rise.

In freshwater, a body will drop at a rate of approximately two feet per second. For example, given a sixty-foot depth, a body will take thirty seconds to reach the bottom. Therefore, in a river with a sixty-foot depth, a body will only be traveling for thirty seconds, or half a minute, once it submerges. If the water is traveling at two knots, in half a minute it will have traveled one hundred feet from the point of submersion, and that is where it will be found on the bottom.

This information can be helpful if you attempt to drag for the body or for later attempts at recovery after the ice roof has melted.

STUDY QUESTIONS

1. Do not submerge your head into an underwater vehicle unless you are _____ and _____.

2. True or false: If a vehicle that has been sitting on ice is about to submerge, you should have the occupants gently open all the doors, if possible.

3. If a vehicle is being held at the surface by the air inside of it, you should _____.

4. Name some of the common potentialities of incidents involving snowmobiles.

5. True or false: No matter how thick the ice, incidents involving ice canopies should be handled by divers wearing helmets.

6. One knot equals a movement of _____ feet per minute.

7. Anything faster than _____ is considered fast-water diving.

8. If the current of a thirty-foot-deep river is traveling at one knot, approximately how far will a body travel from the point of submergence until it reaches the bottom?

12

Ice Formation and Types

Cold air creates cold surface water, and at a given point, the surface begins to freeze. The type of ice produced is determined by many factors, such as wind, salinity, snow, and rain.

How fast did the surface freeze? Was there a subzero cold snap that quickly froze the water, or did the temperature change slowly for a slower freeze? Normally, the faster the water freezes, the stronger the ice will be. Factors such as snow and rain will affect the quality of the ice. If the ice freezes quickly, then the freezing process doesn't have time to reverse itself. If the process occurs slowly, then the surface may go from water to slush, then a little bit back to water, and then back to slush. Areas of hard ice will develop, which may alternately soften and harden. The more the freezing process reverses, the more air will be present in the ice, and the more the ice will expand and contract. Both of these factors will weaken the ice.

Ice is seldom consistent across a body of water. Areas can change in density and type within yards of each other. The safety of the ice depends not only on its thickness, but also on its type, how it froze, how often it has softened and refrozen, whether there is moving or still water below it, wind and weather changes, the age of the ice, and how the ice is used.

How ice is used can have a considerable effect on its integrity. Suppose a car is driving across an ice-covered lake. The ice will be depressed under the weight of the car. As the vehicle continues forward, the ice below it will rise back up again. This creates a wave

139

motion in the water that will follow the car. If one car drives too close behind another, the second car is at increased risk of accidental immersion because of this wave motion, even if the ice is considered sufficiently thick for vehicles.

Picture a large lake with fifty ice fishing shacks in the middle. Ice shacks often weigh more than a hundred pounds, not including the gear and occupants, and some even contain portable heaters. Picture multiple holes in the ice cut for fishing. Why are all of the shacks congregated in this one spot on the lake? They're there because that is where the fish are. Why are the fish there? Perhaps that's where the warm-water springs are. Springs and schooling fish may weaken the ice. Holes cut in the ice will weaken surrounding ice as water runs out onto it.

So, the question isn't whether the ice is good enough to hold the weight of an ice shack, but rather, "Is the spring area of the lake good enough to hold fifty heavy ice shacks with portable heaters and at least fifty ice holes? How do the fishermen reach their shacks? Do they use snowmobiles and cars? Do they perhaps take joyrides around the shacks when they become bored with fishing?"

Is your rescue team fully prepared for a possible incident involving several fisherman in immersed ice shacks, all at least a mile from shore? Is your community prepared to manage an incident with one or more submerged vehicles? If not, your local officials may want to think twice before allowing certain ice activities.

Remember, no matter how thorough you are in checking ice thickness, there is no such thing as ice that is one-hundred percent safe, since you cannot check every inch of it. The more you understand ice and its basic formation, the easier it'll be for you to make an informed decision.

This book isn't meant to cover all types of ice or ice conditions; rather, it's a general reference. More detailed reference materials on ice formation and use can be found elsewhere. The following information is only provided as a starting point.

TYPES OF ICE

Frazzle: Frazzle looks like ice crystals speckled throughout the surface water. This beginning stage of ice can look as if there's an oily film floating on the surface. As the temperature drops, more crystals grow, and so begins the ice. Temperature changes dictate whether it is slow- or quick-forming.

Frazzle slush: This looks like slush and usually forms in areas where the weather or moving water prevents it from freezing solid.

Clear ice: This is the strongest ice, formed by a quick drop in temperature or a long, severe cold spell. It is clear-looking and usually has a black, reflective aspect.

Snow ice: This is milky-looking, weak ice that is basically frozen snow. It is unsafe at any thickness.

Layered ice: This type of ice consists of layers of ice and snow covering more ice and snow. It looks like layered cake. Consider anything less than ten inches to be unsafe.

Drift ice: Unattached to the shore, this type of ice resembles floating islands. If some drifting ice exists, chances are that more will break off.

Pack ice: Pack ice has been pushed up against hard ice by wind or moving water. Weak spots and holes are found where two portions of ice meet.

Polynya: A polynya is an area of open water in ice. A natural opening may be caused by a spring or an upwelling of water, and it will be consistent annually. If so, you can plan for it, marking the danger zone with protective fencing and warning signs.

As a general rule, one to two inches of any type of ice is unsafe.

For comparison purposes, three to four inches of clear ice is safe for one person moving slowly, plus a second person not closer than twenty feet. It can accommodate one fisherman, one ice skater, or one cross-country skier.

Five to six inches of clear ice will allow for two persons moving slowly, with the second person not closer than five feet. It can accommodate two fisherman not closer than ten feet, two skaters at least ten feet apart, or two cross-country skiers not closer than ten feet.

Six to eight inches of clear ice is good for small-group activities, ice skating, a recreational scuba diving operation, or one snowmobile, but no larger vehicles.

Eight to ten inches is acceptable for hockey games and snowmobile races, but no large vehicles, such as automobiles.

Likewise, snow ice and layered ice demand caution and the use of a PFD. Three to four inches of these types is unsafe for *any* activity. Five to six inches of snow ice or layered ice is equivalent to three to four inches of clear ice, being safe for one person moving slowly and a second person not closer than twenty feet away. Six to eight inches is equivalent to five to six inches of clear ice, and eight to ten inches of snow ice or layered ice is sufficient for group activities or one snowmobile but no larger vehicles.

Rain, snow, warm spells, sunshine, cold snaps, wildlife, and human activities can all visibly affect ice on a daily and sometimes hourly basis. When rain stays on ice and doesn't refreeze, it often begins to seep through, causing porosity. When snow falls on the ice, it creates a blanket, resulting in a warming effect. The snow traps the very small amount of heat from the water beneath the ice, causing weakening.

Obviously, warm weather will weaken ice. Water pools from melted ice may form on the surface, causing melting in the same manner as rain. Ironically, even cold conditions can weaken ice. When air temperatures drop, refreezing occurs. The ice then re-expands, which, in turn, causes cracks and density changes.

All types of wildlife affect ice when there is direct contact. Geese are of major concern, since they flock on the ice. Each individual bird creates heat, so a flock close together creates a melting machine. Goose excrement adds to the melting process. Geese can be especially problematic for another reason, since small children often want to get near the birds and chase them. The ice could be weak for as much as fifty feet or more from the outside edge of the gaggle, extending the danger for children.

Schooling fish may also affect the ice above them by their swirling motions and tail movements.

The changing water table can have an extreme effect on the strength, formation, and physical support of the ice. In areas where the water table constantly fluctuates, no use of the ice should be considered. Since there is no water support below the ice, the victim may still be at the surface of the water yet invisible to shore-based rescue personnel.

For the most part, the public doesn't know the difference between good, strong, usable ice and weak, unsupportive ice. Also, people tend to ignore posted safety signs. Telling the public that the ice isn't safe may in itself not be enough. Even fencing off the area may not be enough to ensure that you won't be required to perform an ice rescue in that area.

Educating parents and children can make a big difference in preventing accidents. Explain how no ice is guaranteed to be safe ice. In an ice accident, teach them to call for help first and not to go out onto the ice. Teach them self-rescue techniques. Have rescue rope throw bags stationed at commonly used ice locations.

Teach parents and children about the importance of wearing PFDs when they're on the ice. Even a carbon dioxide or orally inflated PFD can make the difference between life and death in the case of accidental immersion. Go to schools and tell children about the hazards of ice and what to do if they or a friend falls in. Teach children and adults that a person can be more of a hero if he stays onshore and calls for help rather than attempts a freelance rescue. Show them how freelance rescuers hurt an operation more than they help it. Teach them what they can do after they call for help, such as marking the spot on shore in front of the victim, taking ranges, communicating with the victims, attempting to throw a lifesaving aid to the victim, and trying to convince other bystanders to stay onshore as well.

Communities should consider imposing fines and other penalties on adults who irresponsibly allow children to play on unsafe ice, or who themselves become the objects of an ice rescue. If an adult drives a vehicle on the ice with his children, and if the children end up drowning because of an accidental immersion, perhaps mandatory manslaughter laws should be imposed against the adult.

144   STUDY QUESTIONS

1. According to the text, the safety of ice depends not only on its thickness, but also on what other factors? (Name several.)

2. This beginning stage of ice can look as if there's an oily film floating on the surface.

3. The strongest type of ice, formed by a quick drop in temperature or a long, severe cold spell, is called _____.

4. With layered ice, consider any thickness less than _____ to be unsafe.

5. An area of open water in ice is called a _____.

6. In an area where the water table constantly fluctuates, what use of the ice could be considered safe?

13

Cold Stress and Hypothermia

The latter portion of this book has two missions: (1) to teach rescuers how to keep themselves safe and healthy during a winter rescue, and (2) to teach rescuers how to manage patients with cold- and water-related stresses and injuries.

To achieve these objectives, it's necessary to have a basic understanding of heat-loss processes and effects. Certain materials such as the hypothermia temperature chart with core temperatures matched to hypothermia signs and symptoms will not be covered. Few EMS personnel have the capability of taking core temperatures in the field, and even if they did, their prehospital care wouldn't be significantly affected by the resulting information. Instead, they should report the vital signs and other information to the receiving hospital, which hopefully is staffed and equipped to use the information toward providing the most effective treatment.

Instead of discussing core temperatures and matching signs with symptoms, the remainder of this book will focus on knowledge that isn't normally found in standard EMS books. The information presented here can readily be applied to all levels of the rescue operation.

DEFINITIONS

Hypothermia: A cooling of the body's core below normal temperature, often designated as 95°F (35°C) or below.

145

Cold Stress: The direct or indirect effects of heat loss not defined by a specific loss of core temperature. The effects of cold stress can include loss of judgment, confusion, fear, panic, hallucinations, irritability, fatigue, loss of dexterity and strength, numbness, tingling, pain, and uncharacteristic behavior.

Ventricular fibrillation: A condition of the heart involving disorganized, ineffective electrical impulses that prevent the heart from contracting normally. Without defibrillation, this condition usually ends in biological death.

COLD STRESS

When asked to define hypothermia, most who are familiar with the word would answer "the body's inability to maintain normal core temperature" or "loss of core heat." The medical definition of hypothermia is a three-degree core loss, resulting in a body core temperature of less than 95°F. The difference between these two definitions can present a problem for rescuer safety. A person using the medical definition might infer that anything short of a three-degree temperature deficit isn't a serious problem, since it doesn't constitute hypothermia. This is far from correct. How a person feels can be just as important as, if not more so than, the actual quantity of heat lost.

A rescuer without any loss of core temperature might still have such cold hands that he is unable to hold a tether line effectively or save himself from an entanglement. Rescuers who simply feel cold, regardless of core temperature loss, are more likely to take shortcuts, use poor judgment, and rush. An inner voice continually tells them, "I'm cold. I want to get out of here. I want to get back in the warm vehicle. Why am I here?"

Simply feeling cold adds stress to the rescuer's job. A drop in skin temperature results in the release of the hormone epinephrine, which is part of the fight-or-flight mechanism. That release adds to a release already brought on by the rescue incident itself, possibly leading to fear, panic, mental disorganization, and loss of judgment.

The effects of cold and feeling cold can result in physical, mental, and even emotional stress well before medically defined hypothermia occurs. Therefore, this book will operationally define cold stress as the direct or indirect effects of heat loss not defined by a specific core temperature loss, and will not be concerned with core temperatures.

Is a one-degree temperature loss dangerous? *Yes,* even before you lose one degree you can feel cold. If you feel cold, what is going on in your mind? You want to get warm, and you want to hurry the job. If part of your brain is focused on how you feel and a desire to go inside and warm up, then part of your thinking is already diverted from the job at hand. That is the window of opportunity for mistakes to be made.

PATIENT HANDLING

Replay in your mind some of the fire calls and ambulance runs that you have worked on or observed. How many times per call were patients dropped? Hopefully none. How many times were patients yanked up by their arms, banged, and dragged across frozen ground? Again, the answer should be none.

Now, think of the water emergency calls that you have observed or in which you have participated. From the time of being removed from the water or ice and placed in the ambulance, how many times was the victim dropped? How many times was the victim banged into objects such as a boat or backboard? How many times was the victim yanked by the arms or legs and dragged on the ground?

It has been our observation that victims pulled out of the water are dropped, yanked, dragged, or banged *at least* three times before being secured in an ambulance.

Anyone who has experienced a water-related accident is just as much a patient as any victim of a medical or land-based emergency, and he should be treated accordingly. Simple jostling, let alone being dropped, is enough to put a hypothermic heart into cardiac arrest. Immersion hypothermia victims must be handled with extra care, and they certainly cannot afford to be mishandled.

There are several main reasons that water rescue victims are dropped. First, the rescuers may not be trained to work in the water or to handle cold, wet, slippery victims. Departments don't have established protocols for specific water-related emergencies. This often results in chaotic water rescues. Sometimes the incident management system is either not used or it falls apart at water incidents. If the rescuers don't have proper water rescue equipment, they sometimes attempt to make do with equipment designed for land operations. Victims are further jeopardized when the rescuers are cold-stressed, hypothermic, or without proper hand protection.

Rescuers must be prepared for the extremes of exertion common to many water- and ice-related operations.

Many times, the rescuers aren't physically capable of the unexpected amount of exertion required by many water- and ice-related operations. Rescuers often end up in trouble themselves and accidentally mishandle patients while attempting to save themselves. Finally, water-related emergencies are rare compared with other types of calls, so rescuers gain little, if any, practical experience, and departments give water-related incidents last priority for funds and training.

Has your team learned and practiced effective procedures for gently extricating a slippery, helpless, water-laden person out of an ice hole? Have team members practiced working, moving, and handling patients while wearing stiff neoprene gloves and exposure suits, which are probably one-size-fits-few?

Surface ice rescue technicians must be physically fit and capable of crawling hundreds of feet while periodically crashing through the ice. They must be competent swimmers and comfortable in the water. They must be capable of keeping a victim afloat and transporting him to shore. If not, chances are that the victim will either not be reached in time or will be mishandled.

EXPOSURE PROTECTION AND PATIENT HANDLING

Latex EMS gloves and sneakers are not compatible with cold weather. Before you lose one degree of core temperature, your hands could be so vasoconstricted that they have almost no blood in them. Exposing an extremity to cold can result in a loss of blood flow only exceeded by amputation. This translates into a loss of manual dexterity, an inability to walk with a heavy load, and other physiological limitations.

To avoid such consequences, keep duty crew EMS personnel back in the rig where they can stay warm. They shouldn't be allowed to stand outside watching the rescue. If they need to bring the transport device to the hot zone, they should wear warm gloves over their latex gloves. Latex gloves won't protect hands from heat loss to cold air, a wet victim, or a cold metal stretcher. The patient should be handled with warm, strong, hands with full strength, sensation, and dexterity.

EMS personnel and everyone on the scene should wear proper boots to protect their feet from freezing and slipping. They should also wear proper hats, not ball caps, as well as anything else it takes to stay warm.

150 **STUDY QUESTIONS**

1. The onset of hypothermia is often defined as a cooling of the body's core temperature below what temperature?

2. Name some of the effects of cold stress.

3. True or false: Simple jostling, let alone being dropped, is enough to put a hypothermic heart into cardiac arrest.

4. Should duty crew EMS personnel be allowed to participate in the rescue, perhaps as line tenders or spotters? Why or why not?

14

The Four Processes of Heat Loss

Heat is lost from the human body by four basic means: conduction, convection, evaporation, and radiation.

CONDUCTION

Conduction is the transference of heat through solid matter or between objects that are in direct contact. Imagine one warm and one cold object touching each other. Heat will travel from the warm surface to the cold surface until both are the same temperature. For example, your hands will lose heat to cold, wet line, and metal gear.

Water conducts heat away from an object twenty-five times faster than air. This means that your body heat will be ripped out of you far faster in water than in air of the same temperature. Also, water has a tremendous heat capacity. It takes four thousand times more heat to raise the temperature of water one degree than it does to raise air temperature one degree.

Therefore, water will continue to steal your heat for a far longer duration than air will, and it will drop your temperature to a much lower degree because it needs so much more heat for its temperature to be raised. Remember, conduction doesn't stop until both objects are the same temperature.

Conductive Heat Loss and the Rescuer

The body loses heat in 80°F water at the same rate that it does in 42°F air. If you don't picture yourself comfortable working outdoors in your underwear in 42°F air, then you shouldn't consider working in 80°F water without an exposure suit.

The insulation afforded by ¼-inch (5- to 6-mm) neoprene ice rescue suits decreases the ability of water to conduct heat away from your body. It's best if rescuers wear a thick pair of wool socks under the suit, since the boots consist of uninsulated rubber. Keep a pair of thick wool socks along with the extra gloves, hats, boots, and ice awls in your car or rescue vehicle so as always to be prepared. Keep a pair of thick wool socks in the bag of each exposure suit on the rescue vehicle.

Warm gloves with insulation are imperative if you are to prevent rope, metal equipment, wet bodies, and other heat-stealing objects from conducting heat away from your hands, rendering them all but useless. Good boots are important to keep the cold winter ground from ripping heat out of your feet. The human body loses fifteen percent of its heat from the hands and fifteen percent from the feet. This makes insulation of these extremities an imperative. The body loses another twenty-five percent of its heat from the head, particularly the ears, nose, lips, scalp, and face. Rescuers should wear a warm hat that covers the ears and is capable of covering the nose and mouth.

Conductive Heat Loss and the Victim

Never lay out a patient directly on a cold backboard, since it will conduct significant heat from him. Without a blanket, a backboard could conduct more heat away from the patient than the air above. You should secure a wool blanket on the board as insulation under the patient. If wool isn't allowed by your agency while oxygen is being administered (the static electricity common to wool presents a fire hazard), then use another insulating blanket.

If a motor vehicle accident patient is placed on a cold, uninsulated backboard, the patient is likely to shiver. Shivering increases a person's oxygen consumption by fifty percent. Therefore, if you allow a patient to become cold and shiver, he will require fifty percent more oxygen just to make up the difference. If his ability to take in and transport oxygen is already decreased due to trauma, shock, near-drowning, or other conditions, then shivering might be the straw that results in cardiac arrest.

Also, anyone who lacks the ability to move, either because of injury or because of immobilization, will become colder much faster, since he cannot use voluntary muscle movement to generate heat. To make matters worse, the patient experiencing hemorrhagic shock is also much more likely to become hypothermic. An immobilized trauma patient on an uninsulated backboard can be at serious risk of severe shock and even death because of heat loss that can often be prevented.

Also, hypothermia increases bleeding time, which means that a hypothermic trauma patient is at greater risk of experiencing hemorrhagic shock. Hemorrhagic shock will worsen the hypothermia. Rescuers must take as many measures as possible to prevent this vicious cycle.

Avoid using cold, uninsulated backboards. A cardinal rule to remember: insulation over, insulation under, and insulation around. Avoid purchasing sleds, boats, or transport devices that put patients or rescuers in contact with metal. Metal sleds, boats, and ladders steal heat right out of the human body. Also, wet skin can freeze to metal, as any child who has touched his tongue to a frosty doorknob can tell you.

CONVECTION

Your body is surrounded by a cushion of warm air or water heated by conduction. When the wind or water currents carry away this cushion, your body loses heat as it contacts the new surrounding medium. Every time the warm cushion is removed, the body must heat up the new surrounding medium. This transfer of heat by the movement of wind or water is the process of convection.

Convective Heat Loss and the Rescuer

In winter dive calls, your department's SOGs should require taking the time to set up a wind-protected staging area with vehicles, tarps, or other windbreaks. The few minutes this will require are worthwhile and will actually save time in the end. Wind-protected staging areas should be mandatory for all cold-weather, cold-water calls.

Water doesn't have to be fast-moving for convection to occur. Actually, it can be quite still. As soon as you move in water, convection takes place. The trickle of cold water into a wetsuit quickly teaches that lesson. By the process of convection, a victim struggling in cold water will lose heat even more quickly than a still victim will.

Keep the heat-stealing tendencies of water in mind when deciding on personal protective equipment. Properly fitting immersion suits are important, especially in moving water. Proper fit is additionally important, since smaller rescuers are preferred for being on the ice. The one-size-fits-all suits only fit large rescuers. Smaller people in large suits are easily flooded and exhausted.

Convective Heat Loss and the Victim

If the patient must be transported over a long distance, if the wind is blowing, or if the transport device is a hovercraft, an unprotected fast boat, or a helicopter, take extra care to minimize heat loss from convection.

In the case of accidental immersion, don't kick around furiously while waiting for rescuers. Instead, conserve energy by bobbing and huddling. Swimming and working hard in cold water without proper personal protective equipment accelerate death, while bobbing increases survival time in cold water. Bobbing involves minimal movement and energy to stay on the surface. It should be used when there is no ice to hold on to. To bob, take a breath, let your body submerge, and allow it go where it wants to go. Hold your breath for a short period of time but not so long that you feel the need to breathe. Then, exhale as you use a gentle movement of the arms to slowly ascend. Bring only your mouth above the surface, keeping your head supported by the water. Inhale, then repeat these maneuvers. The less you move and the less you raise yourself above the water, the less energy you'll use and the less heat you'll lose.

If a person is buoyant enough, huddling to himself can further decrease heat loss. A person will feel colder bobbing than he will when struggling, but bobbing will result in less heat loss.

EVAPORATION

Evaporation is the process by which a liquid changes to a gas. This change requires significant heat if the liquid is water, since a large quantity of heat is required to raise the temperature of water.

When your body sweats, it is making use of the evaporation process to keep you cool in warm environments and when your core temperature is higher than normal. The perspiration on the skin continues to absorb heat until it is warm enough to change to a vapor. Evaporation from the skin and respiratory tract normally account for twenty percent of total heat loss of the human body. If the skin is wet from an external source, such as rain or immersion, that percentage increases.

Evaporative Heat Loss and the Rescuer

The colder the ambient air temperature, the more body heat is required for skin to dry, which means that it is important to dry yourself off if you exert yourself and perspire in the winter. You may have noticed that, during winter calls, you can actually feel warm. This is because exertion raises the metabolism and shunts warm core blood to the skin, where the cold sensors are. Once the rescue is over and the vehicles are repacked, however, you suddenly feel wet and cold. Before you zip up that jacket, you should dry off your skin, especially your torso and underarms.

Because it is good for repeated uses, a sports chamois is an excellent tool to keep with your jump kit all year round. During one long call, you might need to dry yourself off several times to keep warm. One chamois does the job of many towels and can usually be purchased for under ten dollars. It can be thrown in the washer and dryer when soiled.

If your shirt becomes wet, change it if at all possible. Teams performing strenuous long-distance wilderness rescues should have extra shirts to change into after removing damp undergarments. Rescuers cannot provide optimal patient care when they are cold.

When the air temperature is low or the wind is blowing, evaporation can quickly cause cold stress and hypothermia. It's very important to remove wet exposure suits and gloves as soon as possible. Wet clothing next to your skin functions like a big sweating machine, stealing heat as your body tries to dry itself. As long as your body is wet, you'll be operating at a heat deficit.

Wet feet and hands lose heat more quickly than dry ones, making them more prone to cold injuries, such as frostnip and frostbite. Be sure to wear good, waterproof gloves. Waterproof foot protection is also a must. Make sure your boots are high enough to prevent snow from entering them.

Evaporative Heat Loss and the Victim

Patients pulled from the water are already hypothermic from conduction and convection in the water. When they're pulled out, their heat loss can become even greater by evaporation as they are exposed to wind. To counter this, wrap patients as soon as possible in wool blankets, or put them in a heat-retaining recovery suit.

A heat-retaining recovery suit will minimize evaporative heat loss.

You should only use space blankets as wind protection on the outside of wool blankets. Space blankets work by reflecting heat. If the patient is hypothermic, there's little or no heat to reflect. Worse, space blankets have no insulation or fluid-absorbing capabilities.

RADIATION

In reference to heat transfer, radiation is the emission of infrared energy. Proper insulative exposure equipment will help reduce heat loss due to radiation. Significant heat loss areas include the head, particularly the ears, nose, mouth, and the carotid region of the neck. Other significant areas are the hands, feet, armpits, and groin.

Head protection should cover the ears, lips, nose, and neck. Ears and lips have basically no insulation and therefore radiate much heat. A muffler that can be pulled up over the neck, mouth, nose, and ears is a good idea.

STUDY QUESTIONS

1. Define conduction.

2. Water will conduct heat away from an object _____ times faster than air.

3. It takes _____ times more heat to raise the temperature of water one degree than it does to raise air by the same amount.

4. The human body loses heat in _____°F water at the same rate as it does in 42°F air.

5. The human body loses _____ percent of its heat from the head.

6. Shivering increases a person's oxygen consumption by _____ percent.

7. Does hypothermia increase or decrease bleeding time?

8. Does water have to be fast-moving for convection to occur?

9. Will a victim struggling in cold water lose heat more quickly or less quickly than a victim who remains still?

10. When your body sweats, it is helping to keep itself cool through what process of heat exchange?

15

Heat and Fluid Loss

By both evaporation and convection, the human body loses heat directly from its core simply through the act of breathing. We inhale cold, dry air that is warmed and moistened even before it passes through the trachea. In a few seconds, after circulating it in our lungs, we exhale it again. Every few seconds, our core warms and moistens a new parcel of air, losing heat and water with every breath.

One solution is to wear a muffler to retain some of the exhaled heat and moisture during subsequent inhalations. The side of the muffler against your mouth and nose will become damp with warm, exhaled vapor. The next volume of air will be inhaled through this moisture, hydrating and warming it. A muffler will also decrease heat loss from the nose and lips. The warm moisture will help prevent chapped lips. For those of us working long hours outdoors, this retention can make a big difference in overall comfort and the ability to function. Every once in a while, pull the muffler down to vent any buildup of carbon dioxide.

OXYGEN ADMINISTRATION

A patient who is administered oxygen breathes in cold, very dry gas. Even if the oxygen bottles are kept inside a warm ambulance as opposed to an outside compartment, the gas will still be cold because of the pressure change it experiences as it leaves the cylinder. The decrease in gas pressure from the compressed gas in the cylinder to 14.7 psi ambient sea level pressure can drop the temperature of the oxygen to 22°F. Therefore, it's best to administer warmed, hydrated oxygen to prevent further core heat and fluid loss in the hypothermic patient. Warming can be achieved with simple disposable hydrators with heat packs taped to the outside. Reusable heat packs work well. Reusable baby bottle warmers can also be used.

Another method of warming oxygen during long outdoor transports is to run the oxygen through approximately twenty-five inches of stainless steel tubing coiled in a thermos filled with hot fluid. Although this method won't add moisture to the gas, it allows rescuers to use any available hot fluid. If sterile hot water is available, then run the oxygen directly through the water to hydrate and warm it.

If a demand valve is used to administer oxygen to a breathing patient, or if a positive pressure valve is used to administer oxygen and ventilations to a nonbreathing patient, tape activated heat packs to the end of the intermediate pressure hose closest to the patient, and then wrap the heat packs to insulate them.

For prehospital care, the administration of warmed, moistened oxygen is a method of core rewarming preferable to peripheral, external rewarming. External rewarming dilates peripheral blood vessels, causing core blood to enter cold peripheral tissues. The cold tissues cool the blood, which then returns to the core. This cooling can actually drop the core temperature, putting the patient in rewarming or afterdrop shock.

Additionally, during the rewarming phase is when oxygen is needed most, and it's also the time when oxygen transport to the tissues is most compromised. Therefore, think twice before rewarming external tissues. Check and review local protocols.

ASTHMA

Asthma can be a complication for both rescuers and patients. Asthma has always been an absolute contraindication for sport and public-safety diving, and it should be one for ice and in-water surface rescue as well.

A number of factors can stimulate an asthma attack: inhaling cold, dry air; aspirating water, particularly salt water; increased levels of carbon dioxide; exertion; stress; and simply being cold. Ice and water rescue technicians can easily experience all of these factors simultaneously. Asthmatics can make excellent shore personnel, but they don't belong in the water where they can become a liability to themselves, fellow rescuers, and the original victims.

162   STUDY QUESTIONS

1. By both _____ and _____, the human body loses heat directly from its core simply through the act of breathing.

2. The decrease in pressure when administering oxygen from a compressed-gas cylinder to ambient sea-level pressure can drop the temperature of the oxygen to as low as _____.

3. True or false: For prehospital care, the administration of warmed, moistened oxygen is preferable to external rewarming of the body.

4. According to the text, what are some of the common factors that can stimulate an asthma attack?

16

How Repeated Exposure to Cold Affects Us

Have you ever experienced situations in which you became cold outdoors, went indoors to warm up, went back outside, and discovered that you became colder faster? Each time that you warm up indoors and return outdoors, the interval to the onset of feeling cold decreases. Similarly, did you ever notice that if you quickly warm up by standing next to the intense heat of a fireplace, and if you vasodilate yourself and possibly even perspire, then your ensuing outdoor experience makes you feel colder faster?

Heat loss can occur at an increasing rate. For example, if it took twenty minutes to lose one degree of core temperature, it might only take another ten minutes to lose a second degree of temperature, and an additional five minutes to lose a third. Once hypothermic, it could take twenty-four hours to rewarm to the same physiological state you were in before any heat loss occurred. If you don't wait twenty-four hours before going back outdoors, it could take less time to reach the same chilled state than it did before.

The rate at which the body loses core heat varies from person to person, and it can vary for an individual on a daily basis, depending on hydration, food, fatigue, level of exertion, and other variables. To decrease the rate of heat loss, wear proper exposure protection, including head gear covering the ears, nose, and mouth; wear wool mittens or waterproof insulated gloves, plus proper foot protection; wear loose layers of clothing that can easily be removed to prevent overheating and sweating during exertion; replace

163

164 damp underclothes with dry ones, and dry your skin with a chamois or towel; stay well-hydrated with warm, noncaffeinated, nonalcoholic fluids such as soup, broth, apple cider, or herbal tea; eat healthy meals with complex carbohydrates; get enough sleep; stay dry; and wear proper underclothing for long-duration work. Remember that, although you may feel warmer when exerting yourself in the water, you can lose thirty percent more heat that way than if you just remain still.

Never discount someone's complaints of feeling cold just because you feel warm. Likewise, don't ignore your own feeling of cold if everyone else feels warm. People with little or no body fat will lose heat faster than those with more body fat, and they'll usually feel colder faster. Women typically feel colder faster because their peripheral blood vessels constrict earlier, shunting more blood to the core, resulting in a quicker drop in skin temperature. When skin temperature drops, we feel cold. Women may actually lose less heat than men because the protective vasoconstriction occurs earlier, and they tend to have more body fat and a lower metabolism. Still, as discussed earlier, how rescuers feel may be more important than the actual quantity of heat they lose. How someone feels can greatly affect his behavior and performance.

Men lose more heat because they keep warm blood in peripheral tissues longer and tend to have more muscle mass and, therefore, a higher metabolism. This difference means that a man will feel warmer longer because his skin will be warmer longer. Again, it's important to note that how we feel doesn't necessarily tell us how much heat we're losing.

If you regularly work outdoors, there is another process to consider: burnout. Your brain and body can become so programmed that you actually anticipate feeling cold before you even step outdoors. If you regularly allow yourself to become uncomfortably cold, you can lose enthusiasm and actually feel depressed when a stimulus triggers this anticipation. Although this phenomenon hasn't been formally studied, it can be observed among those who work in the water and routinely become chilled. Scuba instructors, for example, might have a wonderful time with students in the classroom. When they move to the pool and inhale that first whiff of chlorine, however, they begin to lose their enthusiasm. When asked, they often say they wish they didn't have to get wet. Their body and brain already know what is about to happen: Soon they're going to feel cold. Instructors who wear enough thermal protection to remain warm and comfortable during the entire class don't seem to have this problem.

Such burnout also occurs among commercial divers, water-show performers, and in-water researchers. Could you be enthusiastic and energized by something that you know will make you feel cold and uncomfortable? That's when shortcutting really starts happening, and shortcutting on a rescue site can have disastrous results.

It's imperative for the rescuer, therefore, to do anything possible to stay warm and comfortable.

166   STUDY QUESTIONS

1. If you return outdoors after warming yourself by the fire, will you feel cold more quickly or more slowly than you did on your previous outing?

2. According to the text, once hypothermic, it could take how many hours to rewarm yourself to the same physiological state you were in before any heat loss occurred?

3. Although it may make you feel warmer, exerting yourself in water can cause you to lose _____ percent more heat than if you were to remain still.

4. True or false: Men generally lose more heat than women because they keep warm blood in peripheral tissues longer and tend to have more muscle mass and, therefore, a higher metabolism.

17

Cold Stress, Immersion, and Fluid Loss

DEFINITIONS

Ambient medium: The air or water surrounding us.

Cold diuresis: Urine production stimulated by cold.

Dehydration: The loss of body fluids, including blood volume.

Diuresis: Urine production.

Immersion diuresis: Urine production stimulated by immersing the body in a fluid.

Peripheral vasoconstriction: A decrease of the diameter of blood vessels in the outer tissues of the body, thereby sending blood from the peripheral tissues to the core.

Peripheral vasodilation: An increase of the diameter of blood vessels in the outer tissues of the body, sending blood from the core to the peripheral tissues.

Shock: A life-threatening condition of too little oxygen reaching the body's tissues, particularly the vital organs. Since blood is the main carrier of oxygen, shock is often caused by too little blood volume (hypovolemia), which could result from blood loss (hemorrhagic shock), low blood pressure (systolic <80 mm Hg), or a variety of other blood-related causes. A decrease in oxygen can also be caused by respiratory problems such as fluid in the lungs (pulmonary edema) or suffocation (asphyxia). Shock can also be caused by other factors, such as the major emotional stress of seeing one's child drowning (psychogenic shock), which causes a sudden dilation of blood vessels (vasodilation) that severely drops

167

blood pressure. Rewarming or afterdrop shock can occur when hypothermic patients are rewarmed externally, causing vasodilation and a rush of blood to the peripheral tissues, resulting in a drop in blood pressure.

Hydration is an important part of staying warm. Fluid loss typically coincides with heat loss. When you immerse yourself in water, after a few minutes you'll probably feel the urge to urinate. The process involved is called immersion diuresis—urine production stimulated by immersion in a fluid. An understanding of immersion and cold diuresis will help you better take care of yourself, fellow team members, and immersion and hypothermia patients.

When you become cold, the same urge can take place, which is a result of cold diuresis. Perhaps you've noticed that, as you become colder, the urge to urinate increases. A hypothermic patient could be in shock just because of this process.

Immersion in cold water combines the effects of cold and immersion diuresis. Anyone who foolishly attempts a cold water rescue without proper personal protective equipment will be hit by this combination.

PERIPHERAL VASOCONSTRICTION AND VASODILATION

The body's core is usually warmer than the skin. As the blood carries heat from the core to peripheral tissues, it takes core heat with it. The peripheral tissues lose heat to environments colder than the skin. The greater the difference between the temperature of the skin and the environment, the more heat you'll lose.

To decrease heat loss, the body constricts peripheral blood vessels. This keeps warm blood away from the skin, resulting in a drop in skin temperature. Because the skin has less heat, it loses less heat.

On the other hand, when the environment is warm or the core temperature needs to be cooled, peripheral blood vessels dilate, bringing more blood to the skin, where it is cooled by two processes: heat loss to the ambient medium and the evaporation of perspiration. The cooled blood then returns to the core.

On exposure to cold air or water, a series of physiological changes takes place to defend the body and decrease heat loss. These processes affect each other like dominoes in a row:

1. The skin temperature drops toward the ambient temperature.

2. The hormone epinephrine is released. Peripheral blood vessels constrict, sending blood from the peripheral tissues to the warm, insulated core.

3. When blood leaves the peripheral tissues, the temperature of the skin drops.

4. When the skin temperature is lower, it loses less heat to the ambient medium.

Remember the law of conduction: Heat loss will occur until both objects are the same temperature. The greater the difference in temperature between two objects in contact, the faster the warmer object will lose heat to the cooler one. By lowering skin temperature, the body decreases the difference in temperature between itself and the ambient medium. Thus, heat loss is slowed and decreased. This is the principle behind vasoconstriction, keeping warm blood in the insulated core tissues and lowering peripheral tissue temperatures to decrease the amount of heat that the ambient medium can steal.

BLOOD PRESSURE AND DIURESIS

The next step of the cold diuresis process is a rise in blood pressure. Given a state of vasoconstriction, a portion of the blood that used to be in the peripheral tissues is now contained in the core, meaning that the core has more blood volume than usual. This translates into a rise in blood pressure. The colder the ambient medium, the greater the vasoconstriction, the more blood in the core, and the higher the blood pressure. When the body needs to lower blood pressure back to its normal levels, it does so by lowering blood volume. The kidneys respond by pulling water out of the blood, creating urine, and sending it to the bladder, prompting the need to urinate. More accurately, the hypothalamus in the brain stimulates the pituitary gland to suppress antidiuretic hormone (ADH). When ADH is suppressed, the kidneys filter more water out of the blood, creating larger volumes of urine.

The colder the environment, the greater the degree of vasoconstriction; the more peripheral blood sent to the core, the higher the blood pressure; the more the brain tells the kidneys to pull water from the blood, the more urine produced and the lower the blood volume. Even if you refrain from actually urinating, blood volume will still be lost, because once fluid is in the bladder, it cannot be reused by the body.

The body also decreases blood volume by shunting blood fluids from capillaries into the tissues. This space is called the *interstitial space* or *third space*. The fluids remain in the tissues until they are pulled back into the venous capillaries, suffusing back into the blood during rewarming.

The increased blood pressure caused by cold as described above also shows why temporary or chronic hypertension (high blood pressure) is a contraindication to performing ice or water rescues. If a person has high blood pressure to begin with, entering and exerting himself in cold water, plus the effects of dehydration, can raise his blood pressure to dangerous levels.

Team members should have their blood pressure checked before suiting up and after the job is done. Consider a diastolic pressure greater than 100 mm Hg to be the limit for technicians operating in the water or on the ice. Ask the physician presiding over emergency services in your county to designate recommended maximum systolic and diastolic pressures for anyone who wants to serve as an ice or water rescue technician.

As mentioned in Chapter 15, the loss of blood volume from cold diuresis can result in a form of shock called afterdrop or rewarming shock if a hypothermic patient is rewarmed externally. During external rewarming, peripheral blood vessels dilate, sending blood from the core back to the periphery. This can cause a significant drop in blood pressure if the blood volume has already been lowered by cold diuresis. At the same time, the blood that passes through the cold peripheral tissues loses heat. This cold blood then returns to the core, lowering core temperature. A slight rewarming shock is what you may experience when you enter a warm room after becoming cold outdoors. The symptoms are lightheadedness, nausea, and weakness.

Therefore, internal rewarming techniques and procedures that prevent further heat loss are preferred. Examples of such methods include administering prewarmed, hydrated oxygen and placing the patient in a thermal stabilizer capsule. If external warming methods are used, apply heat near core areas, such as the groin and armpits, and gently by the carotid arteries. Do not apply heat directly over the heart, however.

IMMERSION DIURESIS

As mentioned earlier, when we immerse ourselves in water, whether voluntarily or involuntarily, we feel the need to urinate. Immersion diuresis is similar to cold diuresis.

Early Soviet cosmonauts returning from their missions in space experienced severe hypovolemic shock. The condition of weightlessness while in orbit changed the distribution of pressures normally found within their bodies. Under the 1 g influence that we all experience at the earth's surface, the heart has no need to pump blood to the feet, since gravity naturally draws it there. With a decrease in the effect of gravity, as in orbital free-fall, blood pools in the core and blood pressure increases until urine is produced to lower the volume of the blood, thus lowering its pressure.

Although the actual pull of gravity remains unchanged when you are in the water, the effect of water pressure and buoyancy is akin to reducing gravity's effect. In response, the body again pools blood to the core, raising the blood pressure and invoking the kidneys to decrease blood volume by filtering water from the blood.

Pressure is also involved in immersion diuresis, and it's important to understand this for proper patient handling. Water is 800 times denser than air, so it weighs much more and exerts far more pressure. Every foot of the water's depth exerts approximately 0.5 pounds per square inch. A six-foot-tall person hanging vertically in the water has approximately 3 psi more pressure exerted on his feet than on his head. Medical Anti-Shock Trousers (MAST), inflatable trousers placed on patients in shock (systolic blood pressure lower than 80 mm Hg), are designed to create enough pressure on the extremities to keep blood from entering them from the core. The result is that the blood pools in the core, raising the blood pressure. If the MAST are suddenly deflated or removed, the result can be death as blood rushes to the legs, causing the blood pressure to plunge.

MAST are inflated to approximately 100 to 110 mm Hg, which is equivalent to about 2 psi. Therefore, a healthy person hanging vertically in the water will experience pressure on the lower extremities not much less than if he were wearing MAST.

Because of the changes caused by pressure, all hypothermic and immersion patients, regardless of whether or not they have a heartbeat, should be taken out of the water in a horizontal position. They should remain in a horizontal position throughout prehospital care. Extricating a person vertically will result in a rush

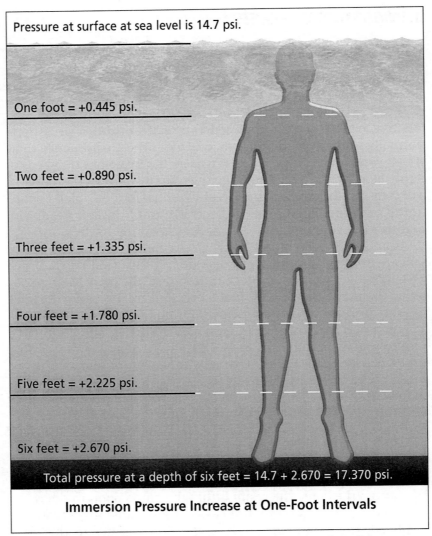

Pressure at surface at sea level is 14.7 psi.

One foot = +0.445 psi.

Two feet = +0.890 psi.

Three feet = +1.335 psi.

Four feet = +1.780 psi.

Five feet = +2.225 psi.

Six feet = +2.670 psi.

Total pressure at a depth of six feet = 14.7 + 2.670 = 17.370 psi.

Immersion Pressure Increase at One-Foot Intervals

A six-foot-tall person hanging vertically in the water has approximately 3 psi more pressure exerted on his feet than on his head.

of blood from the core to the lower extremities. If the person does have a heartbeat, the massive decrease in blood pressure could result in death. If the person is already in cardiac arrest, then the shunting of core blood to the lower extremities can significantly reduce the chances of resuscitation and later survival.

These principles apply to you, the rescuer, as well. If you happen to fall through the ice and then manage to get yourself out,

All victims should be taken out of the water in a horizontal position.

you should continue to roll across the ice until you can be picked up by other rescue personnel. Do not attempt to stand, even if you reach stronger ice or the shore. If you do, blood will rush to your legs, possibly causing you to pass out and fall.

Having some knowledge of such physiological responses is important to the rescuer, both for his own well-being and the victim's safety. If your body fluids and blood volume are decreased due to hypothermia, your ability to function is decreased. When the kidneys filter more water from the bloodstream, the blood becomes more viscous, forcing the heart to work hard and possibly straining it if you overexert yourself or have heart problems. It will also help you recognize the body's responses, such as the light-headedness you feel on entering a warm room after working outside in the cold.

If you warm yourself up and then go back outdoors without drinking water or warm, nonalcoholic, noncaffeinated fluids, then you're asking for trouble. Every time you go in and out of doors to escape the cold, you lose blood volume and with it your ability to think, work, and function safely.

174   DEHYDRATION SIGNS AND SYMPTOMS

If you don't feel the urge to urinate when you're immersed in water, then you are most likely dehydrated.

Dry mouth and thirst are late symptoms of dehydration. Irritability, fatigue, and headaches are indications that you probably need more fluids. Dark urine and pain in the kidney region are late signs and symptoms. The inability to sweat is also a late sign. Muscle cramping can be a symptom of dehydration. If you pinch your skin and it doesn't immediately go back to normal, you are very dehydrated.

If you are well-hydrated, your urine is clear, copious, and frequent.

What happens if you aren't properly hydrated? Imagine this scenario:

You haven't eaten for six hours. You've only drank coffee up until now, and you're tired after a long day at work. On the scene, you realize that you didn't bring your wool hat or gloves. The cold is stinging your skin, but you almost don't feel it because your heart is racing. The media has arrived, along with the victim's family and your chief. The victim is a thirteen-year-old snowmobiler who tried to jump open water and didn't make it. His mother is begging someone to do something to save her son. His mother has been secured by EMS personnel, who are concerned about her heart condition. You can barely hear the boy screaming for help. It's almost dark, so he isn't visible from shore.

The officer in charge at the scene tells you to get dressed as the primary rescuer. You take off your boots to put on the exposure suit, wishing that you'd left a thick pair of wool socks in your car. *Brrr*—that rubber is cold. An EMT comes to record your blood pressure. It's 160/85, probably because of the coffee, the cold, and the adrenaline. You'll be just fine, you tell yourself.

While getting dressed, neither you nor your tender are wearing gloves because you haven't practiced working with them. By the time the suit zipper must be closed, you and your tender's hands are frozen, but because adrenaline is pumping, you both ignore it. You try to close the zipper without taking care to keep loose threads or clothing out of it. Moreover, you forget that closing the zipper is the tender's job. It gets jammed and won't close more than halfway. Now you are cursing and trying to force it up. Everyone is yelling, "What's taking you so long? The kid is screaming!" You lose your temper and the officer in charge comes to tell you to calm down. Luckily, someone standing by with warm, gloved hands and a cool head steps in and gently works the zipper free. You're already breathing heavily, and your heart is racing.

Finally, you and your backup rescuer are dressed and ready to go—in those wonderful, heavy, awkward exposure suits that keep you dry only in theory. Wearing the rescue sling over your shoulder, you crouch low and begin pounding the ice in front of you with an ice pole, working your way across a distance of 250 feet to reach the victim.

At twenty-five feet, you're out of breath. After fifty feet of crouching, banging, and dragging the tether line, you have to stop to open the mouth cover of the suit, which shouldn't have been closed in the first place. You catch your breath. Your leg muscles are cramping and you're in pain. At a hundred feet, your heart feels as if it's pounding out of your chest. You begin asking yourself, "Why am I here?" At that point, the ice breaks and you fall in. You don't reflexively hold the pole out, nor do you cover your face. When your face hits the water, the thermal hammer takes your breath away.

The kid is howling now. He's freezing. Someone on shore is yelling through a megaphone, telling him to hang on and that help will be there any second. All you can think of is how are you going to reach him? With the tender's assistance and your ice picks, you get out of the hole, roll away from it, and begin approaching from a different angle. The suit is wet inside, and your muscles are cramping and aching.

At two hundred feet from shore, you can see the kid with the flashlight attached to your hood, and you call out to him. He is completely unaware of his surroundings and gives no response. Your assessment of him is that he is passive and therefore needs buoyancy immediately. You shout to your backup that you're okay, even though your head is spinning and you want to throw up. You feel weak and light-headed. Another ten feet and you fall in again. You can't do it. You don't feel well at all, and you're on the verge of panicking, yet you maintain an appearance of control. Having also recognized the victim as being passive, the backup moves straight in and effects the rescue.

After finally getting out of the hole, you slowly work your way to shore. Your leaky suit is flooded from being in the water so long, and you're shaking like a leaf. Back in the ambulance, your blood pressure reads 100/60, and you feel terrible mentally and physically. Your brain goes through mental games of what you hoped

176 would have happened, what you should have done, and what went wrong.

Seemingly small actions can stave off such reactions. Eat a sports bar with concentrated complex carbohydrates on the way to the scene (keep a few in your vehicle at all times). Drink warm soup or cider before and after the operation. It is very important to remain well hydrated in the winter. Make sure your urine is clear, copious, and frequent. Have wool gloves, hats, and socks in your vehicle at all times. Know when to tell the IC that you aren't up to operating on the ice and would prefer to act as a tender, if at all. Actions such as these could spell the difference between a rescue and a double recovery.

COFFEE

Avoid coffee at all costs. Coffee is not only bad because it's a diuretic, but also because caffeine is a stimulant. A stimulant will make you think that you're stronger and more alert than you really are. If your last meal was a couple of doughnuts, expect an energy crash, probably just as you set foot on the ice. Coffee can also raise the blood pressure to levels that are unsafe for ice rescue technicians.

ALCOHOL

Alcohol decreases judgment and increases risk taking. It dilates peripheral blood vessels. The drinker feels warmer because core blood rushes out to warm the skin. In reality, more heat may be lost. Alcohol also interferes with the enzymes that allow hemoglobin to carry oxygen in the blood. It increases the chances of failure, injury, and death by reducing the ability to respond appropriately to a threatening situation. It decreases dexterity, increases the chances of vomiting, and interferes with the laryngeal reflex, thereby increasing the chance of drowning. Finally, it is a diuretic, and therefore causes dehydration.

Many water-related rescues happen around 4:00 to 6:00 p.m. Kids come home from school, change their clothes, have a snack, and go out to play. Similarly, adults hit happy hour after work, drink, and sometimes do very foolish, dangerous things. Approximately fifty percent of drownings are alcohol-related.

The potential for a water-related emergency is increased, as is the potential for rescuers themselves who have been drinking. If a team member has consumed any significant quantity of alcohol just prior to the incident, his assignment should be reevaluated.

OTHER WAYS OF LOSING FLUIDS

Breathing

Every time you exhale in cold weather, you see a cloud of vapor. You can lose more than a liter of water in a single day in the winter just by breathing. How many of us drink a liter of water every day? Many of us are slightly dehydrated most of the time. Divers and rescuers who understand the importance of hydration and take in fluids accordingly report less fatigue, fewer headaches, and a general improvement in a feeling of well-being.

Breathing outdoors in cold weather is problematic, robbing our bodies of heat and moisture. To make things worse, the kind of air we breathe indoors during the winter is dry, heated air. A dry mouth means that you aren't properly hydrated. Still, do you become very thirsty when you work outdoors in the wintertime? Probably not. Most of us feel greater thirst in the summer, even though the body does lose copious amounts of moisture during exertion in the winter. In the wintertime, we don't have the same degree of thirst to tell us to drink enough to replace the fluids we've lost. Even if we do drink in the wintertime, we have a tendency to drink more coffee and tea than water.

Shivering

The hypothalamus thermal receptors in the brain are stimulated when skin temperature drops. The hypothalamus causes muscle tensing, otherwise known as shivering. Metabolism can increase three to four times above normal resting values during peaks of shivering. As mentioned earlier, the increase in metabolism produces heat and increases the body's use of oxygen by as much as fifty percent.

Shivering increases heat production, which can increase the core temperature *if* the body is well-insulated. If the body isn't insulated from the environment, or if the environment is particularly conductive, then shivering becomes detrimental.

If an unprotected body is shivering in water, for example, whatever heat is produced will be ripped away, resulting in an even lower core temperature. As fast as the body produces heat by shivering or use of the muscles, cold water can steal it from us.

Remember, the thermal receptors in your skin report how you feel to the hypothalamus in the brain. Your skin, not your core, initially tells you how you feel. Hence, if you work hard in the water, vasodilation will send warm core blood to the peripheral tissues and skin, only making you feel warmer when your core temperature has actually dropped. It is particularly important to prevent trauma patients, or other oxygen-deficient patients, from shivering. The increased oxygen consumption caused by shivering could put hypoxic patients into cardiac arrest. It is vital to keep patients properly warm and dry.

STUDY QUESTIONS

1. Define cold diuresis.

2. Define dehydration.

3. Define peripheral vasoconstriction.

4. Define shock.

5. Can shock be brought on by emotional causes?

6. The colder the ambient medium, the greater the vasoconstriction; the more blood in the core, the higher the _____.

7. When the body needs to reduce blood pressure, it does so by _____.

8. Consider a diastolic pressure greater than _____ to be the limit for technicians operating in the water or on the ice.

9. If a patient is rewarmed externally, the loss of blood volume from cold diuresis can result in a form of shock called _____.

10. MAST are inflated to approximately 100 to 110 mm Hg, which is equivalent to about how many pounds per square inch?

11. Because of the changes caused by pressure, all hypothermic and immersion patients should be taken out of the water in a _____ position.

12. If you don't feel the urge to urinate when you're immersed in water, then you are most likely _____.

13. Coffee is not only bad because it's a diuretic, but also because caffeine is a _____.

14. According to the text, how much water can your body lose in a single day in the winter just by breathing?

15. _____ can increase three to four times above normal resting values during peaks of shivering.

18

Chilblains, Frostnip, Frostbite, and Freezing

Look for cold injuries during the secondary survey of the cold-stressed patient. Areas of the body with frostbite must be handled as gently as a serious fracture.

CHILBLAINS

Chilblains are lesions found on skin that has repeatedly been exposed directly to cool, dry environments of between 32°F and 60°F. People who work outdoors in the winter on a regular basis and who leave their hands, ears, face, and other parts exposed may experience chilblains. Such a lesion is a chronic problem, not something that suddenly arises and disappears for good.

Signs and Symptoms: The lesions show redness and swelling and feel hot to the touch. The affected areas are tender, with itching and sometimes burning sensations.

Treatment: If you aren't sure whether or not you are experiencing chilblains or some other problem, see your physician. If you do have chilblains, do not scratch or further irritate the condition. Protect the lesions, and wear proper exposure garments to prevent reoccurrence or increased severity. If a patient appears to have chilblains, simply protect the areas and make note of them to the receiving hospital.

182 FROSTNIP

Frostnip is the first stage of frostbite. If left untreated, it will become more severe and will affect deeper layers of tissue to become frostbite, which can further progress to freezing. Frostnip isn't a serious problem if treated and if reoccurrence is prevented. Typical locations for it to occur are the ears, nose, cheeks, fingers and other exposed skin areas—yet another reason to wear proper personal protective equipment.

Signs and symptoms: Frostnip is characterized by a redness that later turns white. The affected area lacks sensation. It feels numb to the patient. If you touch an area of skin with frostnip, both the skin and the underlying tissue will feel soft.

Treatment: Treat frostnip by blowing warm air on the affected area or by holding it with a warm hand. Make sure not to expose your warm hand to the cold, so do this treatment indoors if possible, or make sure to insulate the parts of the hand not directly on the frostnip area. Sometimes, the patient may feel burning sensations and tingling during the rewarming process. Explain that this is normal. Blood vessels are dilating back to normal and blood is returning to the area. Nerves are "waking up."

If rewarming doesn't quickly cause a change, then assume the problem to be frostbite, and *stop* the rewarming process.

FROSTBITE

When in doubt as to whether the problem is frostnip or frostbite, assume the worst, frostbite, and do not rewarm the person in the field unless local protocol dictates doing so. Prehospital rewarming of frostbite can lead to severe, permanent tissue damage.

Frostbite generally occurs in unprotected areas that are first affected by peripheral vasoconstriction, such as the ears, nose, face, fingers, hands, toes, and feet. First the skin temperature drops as a result of the blood shift. Without the warmth provided by blood, the fluid in the interstitial space and the cells can literally freeze and form ice crystals. Blood vessels are damaged, further impairing circulation. The skin and underlying subcutaneous tissues are frozen, with little or no oxygen supply.

For a surface ice victim, take care to examine the hands, arms, and whatever other body parts were in continuous contact with the ice. The most likely patients to experience frostbite are those who remain stranded on the ice for a long time.

Signs and symptoms: The frozen skin will appear white and waxy. It will feel cold and hard to the touch. If the underlying tissue isn't frozen, however, the skin will have a resiliency, a bounce. Frostbitten areas may lose sensation.

Treatment: Be very careful not to poke, rub, squeeze, or manipulate frostbitten areas. Such mishandling can result in further tissue damage as ice crystals rupture and tear cells. Handle the area as if it has a serious fracture. Do not attempt to rewarm these areas unless directed by local protocols or a physician at the receiving hospital. Immediately transport the patient to a medical facility. Delays can result in amputations. Cover the affected area with something clean, again taking care not to rub, poke, scratch, shake, put pressure on, or drop frostbitten areas. Do not allow the patient to smoke, which can cause further vasoconstriction and decreased oxygen to the area, or permit him to drink alcohol.

FREEZING

Freezing is frostbite that has progressed to deeper tissues, resulting in the freezing of deeper blood vessels, muscle, bone, organs, and other tissue.

Just as with frostbite, freezing involves ice crystal formation, capillary wall damage, plasma protein molecular changes, cell dehydration, and cell death.

Signs and symptoms: Because both the superficial and underlying tissues are frozen, the area will feel hard to the touch, with no resilience or bounce. The skin color may appear mottled and blotchy, with white, blue grey, and yellow grey areas.

Management: Transport patients to a medical facility as quickly as possible. Handle them as gently as possible. Gently cover affected areas with clean material.

Cold tissues have a lower metabolism than warm tissues, so they need less oxygen. Frozen tissues require even less oxygen. Once rewarming occurs, the need for oxygen increases. When frostbitten areas are rewarmed, their need for oxygen is at its greatest level, yet the ability to deliver oxygen is severely compromised. For this reason, aggressive oxygenation must occur. This procedure is very difficult in the field, especially if the lungs are compromised from aspiration and pulmonary edema. In the hospital, advanced ventilation procedures such as PEEP and CPAP may be used, but even these may not be enough.

184   Hyperbaric oxygen, or oxygen at pressures greater than ambient 14.7 psi, may be helpful during and after the rewarming process. Hyperbaric oxygen improves circulation by increasing the concentration, or partial pressure, of oxygen in the blood; by increasing the flexibility of red blood cells so they can pass through narrower capillaries; by increasing capillary growth; and by decreasing the chances of infection. Hyperbaric oxygen may also treat gangrene problems that can result from frostbite. Discuss this treatment method ahead of time with your hospitals and with the closest hyperbaric facility.

STUDY QUESTIONS

1. Lesions on skin that has repeatedly been exposed directly to cool, dry environments of between 32°F and 60°F are known as _____.

2. Frostnip is characterized by a redness that later turns _____.

3. True or false: You should treat frostnip by blowing warm air on the affected area or by holding it with a warm hand.

4. Prehospital rewarming of frostbite can lead to severe, permanent _____.

5. Name some of the signs and symptoms of frostbite.

6. Handle a frostbite victim as if the frostbitten area has _____.

7. Skin color that appears mottled and blotchy, with white, blue grey, and yellow grey areas, is a sign of _____.

8. Name some of the ways by which hyperbaric oxygen improves circulation.

19

Drowning

A surface incident always has the potential for becoming a sub-surface emergency. For that reason, dive teams should be dispatched immediately. A dive team on the scene can also prevent a victim from submerging if the surface rescuers are delayed in reaching him for whatever reason.

Drowning patients must be handled as gently as possible, just like hypothermic patients. Do not rush! Rushing too often results in the banging, yanking, dragging, and dropping of victims.

DEFINITIONS

Alveoli: Tiny air sacs of the lungs, where the gas exchange between the lungs and pulmonary capillaries takes place. Adults have approximately 400 million alveoli.

Anoxia: A lack of oxygen (an- means without; -oxia refers to oxygen).

Aspiration: The inhalation of foreign materials, such as water or vomit.

Drowning: Death by suffocation after submersion in a liquid. More than 8,000 Americans die annually by drowning. Among accidental causes, it is second only to deaths caused by motor vehicles.

Dry drowning: A drowning with no resulting significant fluid in the lungs, accounting for ten to fifteen percent of all drownings.

187

188   *Hypoxia:* Too little oxygen (hypo- means too little).

Laryngospasm: A sudden, temporary closure of the larynx that can last for up to two minutes. The trachea connects the mouth and nose to the lungs. The esophagus connects the mouth and nose to the stomach. Only one of those tubes can be open at any given time. When we eat or drink, the trachea is closed to prevent airway obstructions. When we breathe, the esophagus is closed to keep air out of the stomach. When water is aspirated and touches the larynx, the body thinks we are about to drink, so it seals off the larynx and trachea with the epiglottis.

Long-term drowning: Submersion for longer than ten minutes.

Near-drowning: Survival for at least twenty-four hours after being saved from submersion or resuscitated from a drowning. If death occurs after twenty-four hours, it is not considered death by drowning. If drowning statistics included those deaths, the numbers reported for fatal drownings would be much higher. Causes of death after twenty-four hours include pneumonia, kidney failure, cerebral edema, adult respiratory distress syndrome (ARDS), and a host of other problems.

Secondary drowning: Pulmonary problems, such as ARDS, after aspiration of fluids during submersion.

Surfactant: The fluid lining of the alveoli that keeps them from collapsing during exhalation.

Wet drowning: Drowning that involves the aspiration of fluid into the lungs, with resultant pulmonary damage. Fluid in the lungs impairs gas exchange and destroys surfactant.

THREE TYPES OF DROWNING

Drownings are classified into three types: dry, wet, and traumatic. Although dry drowning presents the highest chance of a successful rescue if the incident involves a long-term drowning, all drownings should be given equal treatment. The current known record for a successful save of a long-term drowning victim is 88 minutes with no heartbeat. There is always a chance. However, it is important for everyone to understand that no matter how well the rescue is performed, how diligent the prehospital care is, and how extensive the hospital treatment is, there is only a very small chance that the long-term drowning patient will survive. Only ten to fifteen percent of them have even the beginning of a chance.

Do your best, and know you did your best. That is all anyone **189**
can do. Risking your own life won't increase the chances of success, and it might only result in further tragedy.

Dry Drowning

Also known as traumatic asphyxiation, dry drowning is usually
a rapid process lasting seconds or less than a few minutes. The
period of struggle is usually short and not very violent. The victim typically starts out dry and relatively warm, perhaps walking
on the ice, riding in a boat, or standing on a dock. He suddenly
experiences immersion or submersion, accidental or otherwise. At
this point, several dramatic physiological processes occur
simultaneously.

Cold water hits his face and head like a thermal hammer, causing the victim to gasp. If he manages to keep his head out of the
water, the gasping functions as hyperventilation, lowering carbon
dioxide levels. If his face submerges, gasping quickly results in aspiration of water into the trachea, causing a sudden closure of the
epiglottis over the larynx. This is called a laryngospasm. The trachea is now closed for up to two minutes. During that time, the
lungs stay relatively dry. A degree of gas exchange within the body
can take place for up to two minutes after respiration has ceased
if air has been trapped in the lungs by the laryngospasm. Cardiac
arrest follows respiratory arrest, which is caused by asphyxiation.
Cardiac arrest can also occur because of ventricular fibrillation, due
to the effects of very cold water.

The skin's cold receptors trigger the brain to release epinephrine, resulting in massive vasoconstriction and a rush of blood from
peripheral vessels to the core. Cold diuresis and interstitial spac-
ing of fluid is induced, decreasing blood volume. Peripheral tissue
temperature drops dramatically, decreasing cellular metabolism
and the need for oxygen. The decreased need for oxygen decreases
hypoxic cellular death.

Water pressure on the extremities and the decreased effects of
gravity in water result in blood pooling in the core, causing immersion diuresis. The core organs will still be bathed in relatively
warm, well-oxygenated blood.

Basically, the dry drowner is in a state of suspended animation.
The core organs have the warmest, most oxygenated blood. As the
peripheral tissues experience a rapid drop in temperature, their cellular metabolism drops, resulting in a decreased need for oxygen.
This drop is important because the cells aren't receiving a sufficient

supply of oxygen, which is normally accomplished by perfusion, the bathing of the tissues in blood and plasma. The change in metabolism is important also because blood doesn't give up oxygen to cold tissues. The cold, therefore, decreases cellular death in peripheral tissues. The core temperature eventually drops as well, protecting the vital organs from anoxic cellular death.

Children and adults with little body fat have a better chance of surviving a long-term drowning event because of a larger surface-area to body-mass ratio. Their cores will cool more rapidly than those of other body types. A cool core has the same protection as described above for peripheral tissues: The colder the tissue, the less oxygen it needs.

Likewise, the colder the water, the better. Keep in mind, however, that all drowning victims, even toddler bucket-drowning victims, will be hypothermic, so always treat drowning patients as potential long-term drowning saves. It is important to have a standard for time in water to differentiate between rescue and recovery attempts. For this text, the maximum time is somewhere between one and two hours submerged. Anything beyond this time should be treated as a recovery, which means that the operation shouldn't continue until weather conditions are better, daylight is present, and personnel, procedures, and equipment have been reassessed and replenished.

Later resuscitation efforts are aided by several factors. First, relatively dry lungs permit better gas exchange during rescuer ventilation than lungs with fluid in them. A body will incur less cellular death from hypoxia and anoxia because of the decreased metabolism. Finally, it is beneficial to survival that the blood is pooled in the core with the vital organs.

Wet Drowning

Wet drowning is a typical drowning, involving a person who has been in the water for a length of time and who becomes fatigued or for some other reason is unable to stay afloat. The body is already affected by the water with vasoconstriction, a shift of blood volume to the core, immersion and cold diuresis, loss of temperature, and respiratory changes. The person struggles and holds his breath after submergence. Once carbon dioxide tension increases, he reflexively inhales, aspirating water into the lungs. The victim has increased levels of carbon dioxide and lactic acid. The struggling can result in water being ingested into the stomach. A wet drowning is usually a more violent event than a dry drowning. The victim is unlikely to have the suspended animation benefits found in the dry-drowning victim. Consequently, there is less chance that he will survive a long-term drowning.

Traumatic Drowning

This victim is the least likely to be saved. Traumatic drownings are compounded by other problems, ranging from injuries from an automobile accident or water-sports activity to the effects of chemical contamination or intoxication. At any drowning incident, always try to ascertain what additional factors may be involved.

HANDLING THE DROWNING VICTIM

As long as the operation is in the rescue mode, EMS personnel must treat the patient no matter how dead he appears to be. Too often, arriving EMS crews fail to perform optimally because they consider the victim to be dead and beyond resuscitation efforts. Yet it must be a cardinal rule for responders that drowning and hypothermia victims aren't dead until they're warm and dead in the hospital.

When transporting an unconscious victim across open water, keep his airway high above the surface and protected by your hand. A good rescue sling is one that is designed so that the patient's face will fall on the front part of the sling, keeping the airway out of the water as you pull him out of the hole. As long as the rescuer controls the stern of the ice rescue board, the patient's face should never be in the water. Make sure that everyone on the rescue team knows how to properly use whatever transport device they have so as to keep the victim's face as dry as possible. If the transport device is a boat, ramp, or a kayak that isn't self-baling, the rescuers should bring along a baling bucket to keep the vessel dry.

If the patient was in a passive, submerging, or unconscious state during the actual rescue, then responding personnel must take extra care to maintain an airway during transport to shore. During training, have mock victims feign unconsciousness to see what happens to their airways on your transport device. Often transport devices need to be modified.

You should find out why and how the patient fell through the ice. Perhaps the patient experienced a heart attack. Perhaps a snowmobile accident has resulted in trauma. Perhaps a heater inside an ice fishing shack failed and the patient has burn injuries. Compounding injuries may alter the type of handling and treatment administered before and after the patient reaches shore. Always handle a patient as carefully as possible to prevent tissue injury, blood movement to the extremities, and damage to the heart. Maintain the patient in a horizontal position to keep the blood in his core. If you're using inflatable boats, ramps, or kayaks, have roll-up straps in place.

Once a patient reaches shore, perform a primary survey and administer basic life support as dictated. Take a pulse for a full minute on a hypothermic patient before performing CPR. Hyperventilate the patient with sufficient volumes of oxygen. Insufficient ventilations increase the chances of further alveoli collapse and pulmonary edema. Use prewarmed, hydrated oxygen if possible and if local protocols allow it.

Note that, for all drowning types, the presence of blood or foam in the patient's airways is due to alveoli damage and a mixing of aspirated water and bronchial mucous. Proper suctioning is thus important toward maintaining an open airway.

Wrap the patient in blankets from head to toe, over and under. Remember that shivering increases oxygen consumption by fifty percent. Hypothermia increases bleeding time, thereby increasing the chances of hemorrhagic shock. Hemorrhagic shock increases the risk of hypothermia. All of these work together in a vicious cycle that can easily lead to cardiac arrest.

Do not keep the ambulance temperature too high—just keep it high enough that EMS personnel can function comfortably. Remove any wet clothing from the victim and gently dry him off.

If the local protocol for long-term drowning patients involves advanced life support, then follow the protocol, but make sure that you have discussed the pros and cons of ALS with those in charge. Sometimes there is no specific long-term drowning protocol, just a generic drowning protocol. Make sure that everyone has read the latest research and case study papers so that the best protocols are in effect. If defibrillation is part of your local protocol, then make sure the specific protocols for hypothermic and drowning patients are known and used.

Don't forget to perform a secondary survey, since a fall on the ice could result in head, spinal, or other injuries. Treat any freezing, frostbite, and frostnip injuries that may be present. If the victim is conscious, ask him whether someone else may have been involved.

For any number of reasons, it is important to get a victim off the ice as soon as possible.

Possible Head or Spinal Injury

In the event of head or spinal injury, you have to follow the local protocols, but keep in mind that ice is different from the street. The longer victims and rescuers are on the ice, the more likely it is that they will wind up in the water, increasing the risk of drowning. The longer the patient is immobilized in a cold environment, the more heat is lost. The longer technicians work on the ice, the more fatigued and hypothermic they'll become, which will be followed by a decrease in capabilities and judgment. The longer the rescue takes, the more heat the patient will lose, increasing the chances that the heart will go into ventricular fibrillation.

Learning how to perform a field neurological examination is a useful procedure in the ambulance. This examination tests for neurological deficits in greater detail than does a secondary survey. It will provide information to help find out whether the patient experienced a stroke, and it will better describe the degree of hypothermic deficits. If divers are involved, it will help discover whether they have suffered arterial gas embolisms.

ALS for Drowning Victims

Intravenation: It may be very difficult to get an IV into a vaso-constricted, hypothermic patient, which can cause delays in the field.

Most drugs require a certain temperature to work. If the patient is hypothermic, as all ice victims are, then the drugs most likely won't produce the desired effect. Therefore, the paramedic administers another dose, and perhaps even a third dose. When the patient finally undergoes rewarming procedures in the hospital, all of those drugs then take effect.

If rewarming does take place in the field, then the patient's blood volume will simultaneously be increased by the IV fluid and the fluid returning from the tissues, the interstitial space. The result could be pulmonary and cerebral edema.

Ventilations: Intubation is a violent event that could put the hypothermic patient with a weak heartbeat into ventricular fibrillation. Be careful to stimulate the vagal nerve area of the neck as little as possible.

Most ambulances aren't capable of ventilating patients with PEEP or CPAP, so time delayed in the field trying to get an IV in could mean increased problems from anoxia, hypoxia, pulmonary edema, and collapsed alveoli.

Defibrillation: The effects of defibrillation on a very cold heart still aren't fully understood. You should refer to your local protocols for defibrillating hypothermic and drowning patients.

Medical Antishock Trousers: Pulmonary edema is a major contraindication for MAST. Do not use MAST for drowning or near-drowning victims.

Core Temperature		Signs and Symptoms
Degrees F	Degrees C	
98.6°	37°	Sensation of cold. Skin vasoconstriction. Shivering begins.
97°	36°	Shivering increases. Feelings of confusion, with stumbling and disorientation
95°	35°	Impairment of rational thought.
93°	34°	Amnesia. Sensory and motor impairment. Cardiac arrythmias.
91°	33°	Delusions, hallucinations, partial loss of consciousness.
90°	32°	Cyanosis, dilated pupils, respiratory alkalosis. Gross motor impairment.
88°	31°	Shivering stops.
86°	30°	Muscles "gel" and become rigid. Hypoventilation.
84°	29°	Loss of consciousness. Ventricular fibrillation if heart is irritated or victim is dropped.
80°	27°	Appears clinically dead. Muscle rigidity gone. Ventricular fibrillation.
79°	26°	Death likely.

STUDY QUESTIONS

1. The tiny air sacs of the lungs, where the gas exchange between the lungs and pulmonary capillaries takes place, are known as _____.

2. Define anoxia.

3. A sudden, temporary closure of the larynx that can last for up to two minutes is known as a _____.

4. Define long-term drowning.

5. Define near-drowning.

6. Name the three types of drowning.

7. Children and adults with little body fat have a better chance of surviving a long-term drowning event because they have a larger _____ ratio.

8. Is a wet-drowning victim more or less likely to have the suspended-animation benefits of a dry-drowning victim?

9. For how long should you take a pulse on a hypothermic patient before performing CPR?

10. Since a fall on the ice could result in head, spinal, or other injuries, you should always perform a _____.

11. Should MAST be used for drowning or near-drowning victims?

References and Additional Reading

BOOKS

Diving and Subaquatic Medicine. Third Edition. 1992. Eds. Edmonds C., Lowry C., Pennefather, J. *Butterworth Heinemann,* Oxford.

Diving Medicine. Second Edition. 1990. Eds. Bove A., Davis J., *W. B. Saunders Company,* Philadelphia, PA.

Emergency Care. Sixth Edition. 1994. Grant H., Murray R., Bergeron J., *Brady,* Englewood Cliffs, NJ.

Human Performance in the Cold. 1982. Eds. Laursen G., Pozos R., Hempel F. *Undersea Medical Society,* Bethesda, MD.

Hyperbaric Medicine Practice. 1994. Ed. Kindwall, E. *Best Publishing,* Flagstaff, AZ.

Marine Medicine 1993. Proceedings of July 1993 Conference at University of California, San Diego, CA.

Medical Encyclopedia. 1985. Glanze, W., Anderson, K., Anderson, L. *Signet/Mosby,* New York, NY.

198   *NFPA,* Technical Rescue Standards, Document 1670, Jan. 1999.

Polar Diving Workshop. 1991. Eds. Lang M., Stewart J., *American Academy of Underwater Sciences Diving Safety Publication,* Costa Mesa, CA.

Prolonged and Repeated Work in Cold Water. 1985. Ed. Webb, P., *Undersea Medical Society,* Bethesda, MD.

State of Alaska Hypothermia and Cold Water Near-Drowning Guidelines. 1981. Eds. Doolittle W., Haward J., Mills W., Nemiroff M., Samuelson T. *Emergency Medical Services Section, Division of Public Health, Alaska Dept. of Health and Social Services.*

PAPERS AND ARTICLES

Beal M., Richardson E., Brandstetter R., Hedley E., Hochberg F.: Localized brainstem ischemic damage and ondine's curse after near-drowning. *Neurology.* 1983;:33:717-21.

Bennett G.: Hypothermia: management in hospital. *The Journal of Royal Society of Health.* 1988;108, 5:153-4.

Burton H, Coville F., Kocsis C.: Management of submersion hypothermia: successful resuscitation of a 14-year-old girl. *New York State Journal of Medicine.* 1988. August: 434-6.

Downey, R.: The miracle girl. *Lifeguard Systems RDS&R Manual.* 1985.

Gox J., Thomas F., Clemmer T., Grossman M.: A retrospective analysis of air-evacuated hypothermia patients. *Aviation, Space and Environmental Medicine.* 1988; 59:1070-5.

Golden H.: Near-drowning. *Nursing.* 1988; July:33.

Gooden B. A., Lehman R. G., Pym J. *Role of the face in the cardio-vascular responses to total immersion. Aust. J. Exp. Biol. Med. Sci.* 1970; 48:687-90.

Grandey, J.: Hypothermia. *Emergency.* 1995:Jan:28-31.

Hendrick, W.: Executing a river rescue, using roll-up straps. *Fire Engineering*, 1986. Sept., 10-2.

Hendrick, W.: Rescue vs. recovery. *Fire Engineering*, 1986. Sept, 21-3.

Hendrick, W.: Gear up for in-water rescue. *Fire Engineering*, 1986. May, 18-20.

Hendrick, W.: In-water rescue vessels. *Fire Engineering*, 1987. Jan.,43-6.

Hendrick W., Tomson, B.: Hypothermia and the Scuba Diver. *Fire Engineering*. 1987: March 18-20.

Hendrick, W.: The Summer Cold. *Fire Engineering*. 1989.May, 65-7.

Hendrick, W.: Turnout gear in the water, sink or swim. *Fire Engineering*. 1991. July, 78-82.

Hendrick, W.: The parameters of safe ice diving. *Sources*. 1994. March,37-41.

Hendrick, W.: The strong may not survive, part I. *Northeast Dive Journal*. 1995. 5, 21-2.

Hendrick, W.: The strong may not survive, part II. *Northeast Dive Journal*. 1995. October, 21-2.

Hendrick, W.: Technical ice diving. *Northeast Dive Journal*. 1996. Feb, 13-4.

Jacobson, W., Mason L, Briggs B., Schneider S., Thompson J.: Correlation of spontaneous respiration and neurologic damage in near-drowning. *Critical Care Medicine*. 1983; 11, 7:487-9.

Moss, J.: The management of accidental severe hypothermia. *New York State Journal of Medicine*. 1988;88, 8:411-2.

Pearn J.: Pathophysiology of drowning. *The Medical Journal of Australia*. 1985:142:586-8.

200   Schmidt U, Fritz K., Kasperczyk W., Tscherne H.: Successful resuscitation of a child with severe hypothermia after cardiac arrest of 88 minutes. *Prehospital and Disaster Medicine.* 1995;10,1.

Vaughan W.: Distraction effect of cold water on performance of higher-order tasks. *Undersea Biomedical Research* 1977; 4,2:103-16.

Zaferes, A.: Contaminated water—additional safety tips. *Fire Engineering.* 1991. October, 74-5.

Zaferes, A.: Shears cut the need for knives. *ACUC International Contact,* 1996. Spring: 12.

Zaferes, A., Hendrick, W.: What is a field neurological evaluation? 1996. *Sources.* Summer, 42-4.

Zaferes, A., Hendrick, W.: The field neurological examination, part II. *Sources.* 1996. Fall, 38-41.

Zaferes, A.(contributing author): Surface ice rescue. *Technical Rescue.* 1996. Feb.

Answers to Study Questions

Chapter 1

1. No ice is safe.

2. Carry out the rescue as planned.

3. (a) Lack of training, (b) lack of drills or practice, (c) lack of appropriate equipment, (d) responders don't always know what their particular job is at such incidents, (e) since water rescue incidents are less common than other types of incidents, responders don't gain as much practical experience, (f) low budgets, and (g) lack of SOPs or SOGs.

4. Call for help and perform operational, shore-based procedures until help arrives.

5. Proper support.

6. Poor ice. At any ice rescue incident, it has already been proved that the ice is unsafe.

7. (a) Insufficient training, (b) lack of resources, (c) extreme danger, and (d) because he doesn't feel physically or mentally capable of performing the job safely.

8. (a) Bright or clear; (b) dark.

Chapter 2

1. (a) What are the most common problem spots? (b) What are the best access and exit points? (c) How far from shore do the rescues typically take place? (d) What have been the causes of ice incidents in the past? (e) What was the quality of the previous responses? (f) Did the IMS work? (g) What can the community do to prevent ice incidents in the future?

2. (a) To review past ice rescue incidents and find the potential problem areas in your district, (b) to determine the hazards that these areas might present, (c) to determine what is required to make ice operations viable, and (d) to make all personnel and equipment ready for the job.

3. Lipoid pneumonia.

4. EMTs.

5. Alcohol.

6. (1) Behavioral aberrations, including uncooperativeness and belligerency, (2) increased likelihood of risk-taking behavior, (3) the trachea is less likely to close if the victim aspirates water, (4) vomit may cause airway problems, (5) an intoxicated victim may not fully comprehend his predicament, (6) general impairment of reflexes, judgment, coordination, and the senses.

7. The middle.

8. Belaying and tethering techniques.

9. (a) Knowledge of the local bodies of water, (b) the best routes and means of transportation to the site, (c) the personnel and equipment required, and (d) a means of transporting the victim away from the site.

10. SOGs.

Chapter 3

1. Awareness, Operational, Technician.

2. Awareness.

3. On the shore, within ten feet of the water's edge.

4. Technician.

5. True.

6. Warm.

7. The media, family members of the victim, and bystanders.

8. Operational.

Chapter 4

1. The police.

2. Firefighters.

3. (a) Cold stress, (b) immersion hypothermia, (c) near drowning, (d) drowning, (e) long-term drowning, (f) scuba-related injuries.

4. True.

5. The incident commander gives the word and the planned response is known.

6. Provide buoyancy to the victim until the surface crew can reach him.

7. Technician-level training and certification.

Chapter 5

1. The first-arriving officer.

2. His own responsibility.

3. (a) Overlooks the entire operation, (b) is easily seen, and (c) is accessible.

4. True.

5. Information officer.

6. The liaison officer.

7. (a) Yes; (b) yes.

8. True.

Chapter 6

204

1. PFD.

2. III and V.

3. True.

4. (a) Three; (b) 75 feet.

5. Hose, a hose cap, an inflater, a carabiner, and two 150-foot deployment bags.

6. False.

7. 300 to 1,000 feet.

8. Paraffin.

9. Solar plexus.

10. (a) Five; (b) twenty-five.

11. (a) Versatility in a variety of conditions, (b) low profile, (c) stable, (d) easy to right if it is flipped over, (e) lightweight, (f) easily portable, (g) simple to use, (h) nonmental, (i) rounded and foam-protected edges, (j) an aggressive victim may mount it unassisted, (k) compatibility with safe, gentle handling of the victim, including those with spine and head injuries (l) storable in vehicle compartments, (m) affordable, (n) durable, (o) useful for conveying equipment and supporting dive operations, (p) can serve as a tender platform on weak ice.

12. Right an overturned boat in open water.

Chapter 7

1. True.

2. His body language.

3. Reenact exactly what he did immediately prior to and during the incident.

4. Typically they don't lie or create information as adults might.

5. Pinpoint the location of the victim.

6. Twenty.

7. Two feet per second.

8. Spotter.

9. Aggressive, self-rescue (capable), alert, passive, submerging.

10. Self-rescue.

Chapter 8

1. (a) Passive victims, (b) submerging victims, and (c) victims experiencing trauma or other medical emergencies.

2. Reach method.

3. Buoyant and nonmetal.

4. Pull the far end of the hose around the victim, forming a U.

5. (a) Hitting the victim with a tire might injure him or cause him to go under, (b) a tire could damage the ice, (c) a tire could knock a victim away from his handhold, and (d) tires float high in the water and are difficult for weak, hypothermic victims to grasp.

6. (a) Screw; (b) pound.

Chapter 9

1. True.

2. Low and slow.

3. Twenty.

4. No.

5. One foot per second.

6. Egress.

7. (a) By breaking the supportive ice, the victim is put at much greater risk of submersion, (b) if the victim goes down, his location might be lost, and (c) breaking the ice will nullify the option of using it as a platform for operations.

8. Ten seconds.

9. The side.

10. False.

11. Wrist.

12. (a) Circular; (b) two to three feet in diameter.

Chapter 10

1. (a) Tension the line and slowly pull, (b) slacken the line, and (c) stop.

2. Stop.

3. Arm straight up with a fist.

4. One pull.

5. Wave arm up and down vertically.

6. Two blasts.

7. Arm up, making large circular motions.

8. Three blasts.

9. Arm up, sweeping side to side over the head.

10. Four pulls.

Chapter 11

1. (a) A certified ice diving rescue technician using scuba; (b) trained in underwater vehicle extrication.

2. True.

3. Stay away from it and initiate dive operations.

4. (a) The incident sites are often in remote locations, (b) snowmobilers are often farther out from shore than skaters and walkers, (c) alcohol is frequently involved, (d) snowmobilers are more likely to have compounding trauma problems, (e) their helmets and heavy boots can complicate matters for rescue and EMS personnel, and (f) fuel from the machine can pose a hazard.

5. True.

6. One hundred.

7. One knot.

8. Twenty-five feet.

Chapter 12

1. (a) Its type, (b) how it froze, (c) how often it softened and refroze, (d) whether the water below it is moving or still, (e) wind and weather changes, (f) the age of the ice, and (g) how the ice is used.

2. Frazzle.

3. Clear ice.

4. Ten inches.

5. Polynya.

6. No use of the ice should be considered safe.

Chapter 13

1. 95°F.

2. Loss of judgment, confusion, fear, panic, hallucinations, irritability, fatigue, loss of dexterity and strength, numbness, tingling, pain, and uncharacteristic behavior.

3. True.

4. No. EMS personnel should stay warm so as to maintain their manual dexterity, as well as for other physiological reasons.

Chapter 14

1. Conduction is the transference of heat through solid matter or between objects that are in direct contact.

2. Twenty-five times faster.

3. 4,000 times more heat.

4. 80°F.

5. Twenty-five percent.

6. Fifty percent.

7. Increase.

8. No.

9. More quickly.

10. Evaporation.

Chapter 15

1. Evaporation and convection.

2. 22°F.

3. True.

4. (a) Inhaling cold, dry air, (b) aspirating water, particularly salt water, (c) increased levels of carbon dioxide, (d) exertion, (e) stress, and (f) being cold.

Chapter 16

1. More quickly.

2. Twenty-four hours.

3. Thirty percent.

4. True.

Chapter 17

1. The production of urine as stimulated by the cold.

2. The loss of body fluids, including blood volume.

3. A decrease of the diameter of blood vessels in the outer tissues of the body, thereby sending blood to the core.

4. A life-threatening condition of too little oxygen reaching the body's tissues, particularly the vital organs.

5. Yes.

6. Blood pressure.

7. Drawing water out of the blood (creating urine).

8. 100 mm Hg.

9. Afterdrop shock.

10. 2 psi.

11. Horizontal.

12. Dehydrated.

13. Stimulant.

14. More than a liter.

15. Metabolism.

Chapter 18

1. Chilblains.

2. White.

3. True.

4. Tissue damage.

5. (a) The frozen skin will appear white and waxy, (b) the skin will feel cold and hard to the touch, (c) if the underlying tissue isn't frozen, the skin will have a resiliency, and (d) frostbitten areas may lose sensation.

6. A serious fracture.

7. Freezing.

8. (a) By increasing the concentration of oxygen in the blood, (b) by increasing the flexibility of red blood cells so they can pass through narrower capillaries, (c) by increasing capillary growth, and (d) by decreasing the chances of infection.

210   **Chapter 19**

1. Alveoli.

2. A lack of oxygen.

3. Laryngospasm.

4. Submersion for longer than ten minutes.

5. Survival for at least twenty-four hours after being saved from submersion or resuscitated from a drowning.

6. Dry drowning, wet drowning, and traumatic drowning.

7. Surface-area to body-mass ratio.

8. Less likely.

9. One minute.

10. Secondary survey.

11. No.

Index